Jaguars and Electric Eels

Venezuela, 1800

ALEXANDER VON HUMBOLDT

Jaguars and Electric Eels

Translated by JASON WILSON

GREAT
JOURNEYS

PENGUIN BOOKS

Published by the Penguin Group
Penguin Books Ltd, 80 Strand, London WC2R ORL, England
Penguin Group (USA) Inc., 375 Hudson Street, New York, New York 10014, USA
Penguin Group (Canada), 90 Eglinton Avenue East, Suite 700, Toronto, Ontario, Canada M4P 2Y3
(a division of Pearson Penguin Canada Inc.)
Penguin Ireland, 25 St Stephen's Green, Dublin 2, Ireland (a division of Penguin Books Ltd)
Penguin Group (Australia), 250 Camberwell Road, Camberwell, Victoria 3124, Australia
(a division of Pearson Australia Group Pty Ltd)
Penguin Books India Pvt Ltd, 11 Community Centre, Panchsheel Park, New Delhi – 110 017, India
Penguin Group (NZ), 67 Apollo Drive, Rosedale, North Shore 0632, New Zealand
(a division of Pearson New Zealand Ltd)
Penguin Books (South Africa) (Pty) Ltd, 24 Sturdee Avenue, Rosebank, Johannesburg 2196, South Africa

Penguin Books Ltd, Registered Offices: 80 Strand, London WC2R ORL, England

www.penguin.com

Personal Narrative of a Journey to the Equinoctial Regions of the New Continent first
published in Penguin Classics 1995
This extract published in Penguin Books 2007

1

Translation copyright © Jason Wilson, 1995
All rights reserved

The moral right of the translator has been asserted

Inside-cover maps by Jeff Edwards

Taken from the Penguin Classics edition of *Personal Narrative of a Journey to the Equinoctial
Regions of the New Continent*, translated by Jason Wilson

Typeset by Rowland Phototypesetting Ltd, Bury St Edmunds, Suffolk
Printed in England by Clays Ltd, St Ives plc

ISBN: 978-0-141-03206-1

Contents

Alexander von Humboldt (1769–1859) set off on his journey around the New World in 1799, spending five years there. He wrote his great account, *Relation historique du voyage aux régions équinoxiales du nouveau continent*, in thirty volumes. The last three volumes, from which this book is drawn, comprised his *Personal Narrative*, an overwhelmingly engaging picture of a great scientific and cultural mind reacting to Tropical America. These extracts focus on von Humboldt's experiences travelling from the Venezuelan coast to the banks of the Río Negro.

From my earliest days I felt the urge to travel to distant lands seldom visited by Europeans. This urge characterizes a moment when our life seems to open before us like a limitless horizon in which nothing attracts us more than intense mental thrills and images of positive danger. I was brought up in a country that has no relations with either of the Indies, and I lived in mountains far from the sea and famous for their working mines, yet I felt an increasing passion for the sea and a yearning to travel far overseas. What we glean from travellers' vivid descriptions has a special charm; whatever is far off and suggestive excites our imagination; such pleasures tempt us far more than anything we may daily experience in the narrow circle of sedentary life. My taste for botanizing and the study of geology, with the chance of a trip to Holland, England and France accompanied by Georg Forster, who was lucky enough to travel with Captain Cook on his second world tour, helped determine the travel plans I had been hatching since I was eighteen years old. What attracted me about the torrid zone was no longer the promise of a wandering life full of adventures, but a desire to see with my own eyes a grand, wild nature rich in every conceivable natural product, and the prospect of collecting facts that might contribute to the progress of science.

[...]

Journey from Cumaná to La Guaira – The road to Caracas – General observations on the provinces of Venezuela – Caracas

Crossing from Cumaná to La Guaira by sea our plan was to stay in Caracas until the end of the rainy season; from there we would go to the great plains, the llanos, and the Orinoco missions; then we would travel upstream on the great river from south of the cataracts to the Río Negro and the Brazilian frontier, and return to Cumaná through the capital of Spanish Guiana, called Angostura [now Ciudad Bolívar] or Straits. It was impossible to calculate how long this journey of some 700 leagues would take in canoes. On the coasts only the mouth of the Orinoco is known. No trading is carried out with the missions. What lies beyond the plains is unknown country for the inhabitants of Caracas and Cumaná. In a land where few travel, people enjoy exaggerating the dangers arising from the climate, animals and wild men.

The boat that took us from Cumaná to La Guaira was one of those that trade between the coasts and the West Indies Islands. They are 30 feet long, and not more than 3 feet above the water, without decks. Although the sea is extremely rough from Cape Codera to La Guaira, and although these boats have large triangular sails, not one of them has been lost at sea in a storm. The skill of the Guaiquerí pilots is such that voyages of 120 to 150 leagues in open sea, out of sight of land, are done without charts or compasses, as with the ancients. The Indian pilot guides himself by the polar star or the sun.

When we left the Cumaná coast we felt as if we had been living there for a long time. It was the first land that we had reached in a world that I had longed to know from my childhood. The impression produced by nature in the New World is so powerful and magnificent that after only a few months in these places you feel you have been here years. In the Tropics everything in nature seems new and marvellous. In the open plains and tangled jungles all memories of Europe are virtually effaced as it is nature that determines the character of a country. How memorable the first new country you land at continues to be all your life! In my imagination I still see Cumaná and its dusty ground more intensely than all the marvels of the Andes.

As we approached the shoal surrounding Cape Arenas we admired the phosphorescence of the sea. Bands of dolphins enjoyed following our boat. When they broke the surface of the water with their broad tails they diffused a brilliant light that seemed like flames coming from the depths of the ocean. We found ourselves at midnight between some barren, rocky islands in the middle of the sea, forming the Caracas and Chimanas groups. The moon lit up these jagged, fantastic rocks, which had not a trace of vegetation. All these islands are uninhabited, except one where large, fast, brown goats can be found. Our Indian pilot said they tasted delicious. Thirty years back a family of whites settled here and grew maize and cassava. The father outlived his children. As he had become rich he bought two black slaves, who murdered him. Thus the goats ran wild, but not the maize. Maize appears to survive only if looked after by man. Birds destroy all the seeds needed to reproduce. The two slaves escaped punishment, as nothing could be proved. One of the blacks is now the hangman at Cumaná. He betrayed his companion, and obtained pardon by accepting being hangman.

We landed on the right bank of the Neveri and climbed to the little fort of El Morro de Barcelona, built some 60 to 70 *toises* above sea-level. We remained five hours in this fort guarded by the provincial militia. We waited in vain for news about English pirates stationed along the coast. Two of our fellow travellers, brothers of the Marquis of Toro in Caracas, came from Spain. They were highly cultivated men returning home after years abroad. They had more reason to fear being captured and taken as prisoners to Jamaica. I had no passport from the Admiralty, but I felt safe in the protection given by the English Government to those who travel for the progress of science.

The shock of the waves was felt in our boat. My fellow travellers all suffered. I slept calmly, being lucky never to suffer seasickness. By sunrise of the 20th of November we expected to double the cape in a few hours. We hoped to arrive that day at La Guaira, but our Indian pilot was scared of pirates. He preferred to make for land and wait in the little harbour of Higuerote until night. We found neither a village nor a farm but two or three huts inhabited by mestizo fishermen with extremely thin children, which told us how unhealthy and feverish this coast was. The sea was so shallow that we had to wade ashore. The jungle came right down to the beach, covered in thickets of mangrove. On landing we smelled a sickly smell, which reminded me of deserted mines.

Wherever mangroves grow on the seashore thousands of molluscs and insects thrive. These animals love shade and half light, and in the scaffolding of the thick intertwined roots find shelter from the crashing waves, riding above the water. Shellfish cling to the network of roots; crabs dig into the hollow trunks, and seaweeds, drifting ashore, hang from branches and bend them down. Thus, as the mud accumulates between the roots, so dry land moves further and further out from the jungly shores.

When we reached the high seas my travelling companions got so scared from the boat's rolling in a rough sea that they decided to continue by land from Higuerote to Caracas, despite having to cross a wild and humid country in constant rain and flooding rivers. Bonpland also chose the land way, which pleased me as he collected numerous new plants. I stayed alone with the Guaiquerí pilot as I thought it too dangerous to lose sight of the precious instruments that I wanted to take up the Orinoco.

La Guaira is more a bay than a harbour; the sea is always rough, and boats are exposed to dangerous winds, sandbanks and mist. Disembarking is very difficult as large waves prevent mules from being taken ashore. The negroes and freed mulattos who carry the goods on to the boats are exceptionally muscular. They wade into the water up to their waists and, surprisingly, are not scared of the sharks that teem in the harbour. The sharks are dangerous and bloodthirsty at the island opposite the coast of Caracas, although they do not attack anybody swimming in the harbour. To explain physical phenomena simply people have always resorted to marvels, insisting that here a bishop had blessed the sharks in the port.

We suffered much from the heat, increased by the reverberation from the dry, dusty ground. However, the excessive effect of the sun held no harmful consequences for us. At La Guaira sunstroke and its effects on the brain are feared, especially when yellow fever is beginning to appear. One day I was on the roof of our house observing the meridian point and the temperature difference between the sun and shade when a man came running towards me and begged me to take a drink he had brought along with him. He was a doctor who had been watching me for half an hour out in the sun from his window, without a hat on my head, exposed to the sun's rays. He assured me that coming from northern climes such imprudence would undoubtedly lead that night to an attack of yellow fever if I did not take his medicine. His prediction, however seriously argued, did not alarm me as I had had plenty of time to get acclimatized. But how could I refuse his argument when he was so polite and caring? I swallowed his potion, and the doctor must now have included me in the list of people he had saved from fever that year.

From La Venta the road to Caracas rises another 150 toises to El Guayabo, the highest point; but I continued to use the barometer until we reached the small fort of Cuchilla. As I did not have a pass – for over five years I only needed it once, when I first disembarked – I was nearly arrested at an artillery post. To placate the angry soldiers I transformed the height of the mountains into Spanish *varas*. They were not particularly interested in this, and if I had anyone to thank for my release it was an Andalusian who became very friendly the moment I told him that the Sierra Nevada of his home were far higher than any of the mountains around Caracas.

When I first travelled the high plateaux towards Caracas I met many travellers resting mules at the small inn of Guayabo. They lived in Caracas, and were arguing over the uprising that had recently taken place concerning the independence of the country. Joseph España had died on the scaffold. The excitement and bitterness of these people, who should have agreed on such questions, surprised me. While they argued about the hate mulattos have for freed blacks, about the wealth of monks, and the difficulties of owning slaves, a cold wind, which seemed to blow down from La Silla, enveloped us in a thick mist and ended the animated discussion. Once inside the inn, an old man who before had spoken with great equanimity, said to the others that it was unwise to deal with political matters at a time when spies could be lurking around, as much in the mountains as in the cities. These words, spoken in the emptiness of the sierra, deeply impressed me; I was to hear them often during our journeys.

Caracas is the capital of a country almost twice the size of Peru and only a little smaller than Nueva Granada (Colombia). This country is officially called in Spanish the Capitanía-General de Caracas or the Capitanía-General de las Provincias de Venezuela, and has nearly a million inhabitants, of whom some 60,000 are slaves. The copper-coloured natives, the *indios*, form a large part of the population only where Spaniards found complex urban societies already established. In the Capitanía-General the rural Indian population in the cultivated areas outside the missions is insignificant. In 1800 I calculated that the Indian population was about 90,000, which is one ninth of the total population, while in Mexico it rose to almost 50 per cent.

Among the races making up the Venezuelan population blacks are important – seen both compassionately for their wretched state, and with fear due to possible violent uprisings – because they are concentrated in limited areas, not so much because of their total number. Of the 60,000 slaves in the Venezuelan provinces, 40,000 live in the province of Caracas. In the plains there are only some 4,000 to 5,000, spread around the haciendas and looking after the cattle. The number of freed slaves is very high as Spanish legislation and custom favour emancipation. A slave-owner cannot deny a slave his freedom if he can pay 300 piastres, even if this would have cost the slave-owner double because of the amount of work the slave might have done.

After the blacks I was interested in the number of white *criollos*, who I call Hispano-Americans, and those whites born in Europe. It is difficult to find exact figures for such a delicate issue. People in the New World, as in the Old, hate population censuses because they think they are being carried out to increase taxation. The number of white *criollos* may reach some 200,000 to 210,000 people.

I remained two months in Caracas. Bonpland and I lived in a large virtually isolated house in the elevated part of the city. From the gallery we could see the La Silla peak, the serrated crest of the Galipano, and the cheerful Guaire valley whose leafy fields contrasted with the curtain of the mountains around. It was the dry season. To improve the land the savannah and grass on the rocks were set on fire. Seen from far off, these great fires created surprising light effects. Wherever the savannah climbed up the slopes and filled the gorges cut by torrential waters these strips of land on fire seemed at night like lava hanging above the valley.

If we had reasons to be pleased with the location of our house we had even more for the way we were welcomed by people from all classes. I have had the advantage, which few Spaniards can share with me, of having successively visited Caracas, Havana, Bogotá, Quito, Lima and Mexico, and of making contact with men of all ranks in these six capitals. In Mexico and Bogotá it seemed to me that interest in serious scientific studies predominated; in Quito and Lima people seemed more inclined to literature and all that flatters a lively imagination; in Havana and Caracas, there predominated a broader culture in political matters, more open criteria about the state of the colonies and metropolis. Intense commerce with Europe and the Caribbean Sea have powerfully influenced the social evolution of Cuba and the beautiful provinces of Venezuela. Nowhere else in Spanish America does civilization appear so European.

In the colonies skin colour is the real badge of nobility. In Mexico as well as in Peru, at Caracas as in Cuba, a barefoot man with a white skin is often heard to say: 'Does that rich person think himself whiter than I am?' Because Europe pours so many people into America, it can easily be seen that the axiom 'Todo blanco es caballero' (All whites are gentlemen) must wound the pretensions of many ancient and aristocratic European families. We do not find among the people of Spanish origin that cold and pretentious air which modern civilization has made more common in Europe than in Spain. Conviviality, candour and great simplicity of manner unite the different classes in the colonies.

In several families I found a feeling for culture. They know about the great works of French and Italian literature; music pleases them, and is played with talent, which like all of the arts unites the different social classes. The exact sciences, and drawing and painting, are not as well established here as they are in Mexico and Bogotá, thanks to the liberality of the government and the patriotism of the Spanish people.

In a country with such ravishing views I hoped to find many people who might know about the high mountains in the region; and yet we could not find one person who had climbed to La Silla's peak. Hunters do not climb high enough, and in these countries nobody would dream of going out to look for alpine plants, or to study rock strata, or take barometers up to high altitudes. They are used to a dull domestic life, and avoid fatigue and sudden changes in climate as if they live not to enjoy life but to prolong it.

The Captain-General, Sr Guevara, lent us guides; they were negroes who knew the way that led to the coast along the sierra ridge near the western peak. It is the path used by smugglers, but neither our guides nor the most experienced militia, formed to chase the clandestine traffickers, had ever climbed to the eastern La Silla peak.

We set off before sunrise, at five in the morning, with the slaves carrying our instruments. Our party consisted of eighteen people, and we advanced in Indian file along a narrow path on a steep grassy slope. From La Puerta the path becomes steep. You have to lean forward to climb. The thick grass was very slippery because of the prolonged drought. Cramp-irons and iron-tipped sticks would have been very useful. Short grass covers the gneiss rocks; it is impossible to grip it or dig steps into it as in softer soil. More tiring than dangerous, the climb soon disheartened the men accompanying us who were not used to mountain climbing. We wasted a lot of time waiting for them, and did not decide to continue alone until we saw them returning down the mountain instead of climbing up after us. Bonpland and I foresaw that we would soon be covered in thick fog. Fearing that our guides would use the fog to abandon us we made those carrying the instruments go ahead of us. The familiar chatting of the negroes contrasted with the taciturn seriousness of the Indians who had accompanied us up to then. They joked about those who had spent hours preparing for the ascent, and then abandoned it straightaway.

After four hours walking through savannah we reached a little wood composed of shrubs called *el pejual*, perhaps because of the amount of *pejoa* (*Gaultheria odorata*) there, a plant with strong-smelling leaves. The mountain slope became more gentle and we could pleasurably study the plants of the region. Perhaps nowhere else can so many beautiful and useful plants be discovered in such a small space. At 1,000 toises high the raised plains of La Silla gave place to a zone of shrubs that reminded one of the *páramos* and *punas*.

Even when nature does not produce the same species in analogous climates, either in the plains of isothermal parallels or on tablelands whose temperature resembles that of places nearer the poles, we still noticed a striking resemblance of appearance and physiognomy in the vegetation of the most distant countries. This phenomenon is one of the most curious in the history of organic forms. I say history, for reason cannot stop man forming hypotheses on the origin of things; he will always puzzle himself with insoluble problems relating to the distribution of beings.

A grass from Switzerland grows on the granitic rocks of the Magellan Strait. New Holland contains more than forty European phanerogamous plants. The greater amount of these plants, found equally in the temperate zones of both hemispheres, are completely absent in the intermediary or equinoctial regions, on plains and on mountains. A hairy-leafed violet, which signifies the last of the phanerogamous plants on Tenerife, and long thought specific to that island, can be seen 300 leagues further north near the snowy Pyrenean peaks. Grasses and sedges of Germany, Arabia and Senegal have been recognized among plants collected by Bonpland and myself on the cold Mexican tablelands, on the burning Orinoco banks and on the Andes, and at Quito in the Southern hemisphere. How can one believe that plants migrate over regions covered by sea? How have the germs of life, identical in appearance and in internal structure, developed at unequal distances from the poles and from the oceans, in places that share similar temperatures? Despite the influence of air pressure on the plants' vital functions, and despite the greater or lesser degree of light, it is heat, unequally distributed in different seasons, that must be considered vegetation's most powerful stimulus.

The amount of identical species in the two continents and in the two hemispheres is far less than early travellers once led us to think. The high mountains of equinoctial America have their plantains, valerians, arenarias, ranunculuses, medlars, oaks and pines, which from their features we could confuse with European ones, but they are all specifically different. When nature does not present the same species, she repeats the same genera. Neighbouring species are often found at enormous distances from each other, in low regions of a temperate zone, and on mountains on the equator. And, as we found on La Silla at Caracas, they are not the European genera that have colonized mountains of the torrid zone, but genera of the same tribe, which have taken their place and are hard to distinguish.

The more we study the distribution of organized life on the globe, the more we tend to abandon the hypothesis of migration. The Andes chain divides the whole of South America into two unequal longitudinal parts. At the foot of this chain, on both east and west, we found many plants that were specifically identical. The various passes on the Andes would not let any vegetation from warm regions cross from the Pacific coast to the Amazon banks. When a peak reaches a great height, whether in the middle of low mountains and plains, or in the centre of an archipelago raised by volcanic fires, its summit is covered with alpine plants, many of which are also found at immense distances on other mountains under similar climates. Such are the general phenomena of plant distribution.

There is a saying that a mountain is high enough to reach the rhododendron and befaria limit, in the same way one says one has reached the snow limit. In employing this expression it is tacitly assumed that under identical temperatures a certain kind of vegetation must grow. This is not strictly true. The pines of Mexico are absent in the Peruvian Andes. The Caracas La Silla is not covered with the same oaks that flourish in New

Granada at the same height. Identity of forms suggests an analogy of climate, but in similar climates the species may be very diversified.

The attractive Andean rhododendron, or befaria, was first observed by Mutis near Pamplona and Bogotá, in the 4th and 7th degree of latitude. It was so little known before our expedition up La Silla that it was not to be found in any European herbal. The learned editors of *The Flora of Peru* had even described it under another name. The two species of befaria we brought down from La Silla are specifically different from those at Pamplona and Bogotá. Near the equator the Andean rhododendrons cover the mountains right up to 1,600 and 1,700 toises. Going further north on La Silla we find them lower, below 1,000 toises. Befaria recently discovered in Florida, in latitude 30, grow on low hills. Thus, within 600 leagues in latitude, these shrubs descend towards the plains in proportion as their distance from the equator increases.

Due to the thickness of the vegetation, made up of a plant of the Musaceae family, it was hard to find a path. We had to make one through that jungle of musaceous plants; the negroes led us, cutting a path with machetes. We saw the peak at intervals through breaks in the cloud, but soon we were covered in a thick mist and could only proceed using the compass; with each step we risked finding ourselves at the edge of a precipice, which fell 6,000 feet down to the sea. We had to stop, surrounded by cloud down to the ground, and we began to doubt if we would reach the eastern peak before sunset. Luckily the negroes carrying the water and the food had arrived, so we decided to eat something. But the meal did not last long because either the Capuchin father had not calculated our numbers properly or the slaves had already eaten everything. We found only olives and some bread. We had been walking for nine hours without stopping or finding water. Our guides seemed to lose heart, and wanted to go back. Bonpland and I had difficulty in persuading them to stay with us.

To reach the peak we had to approach as near as possible to the great cliff that falls to the coast. We needed three quarters of an hour to reach the top. While sitting on the peak observing the inclination of the magnetic needle I saw a great number of hairy bees, somewhat smaller than the northern European ones, crawling all over my hands. These bees nest in the ground and rarely fly. Their apathy seemed to derive from the cold mountain air. Here they are called *angelitos* (little angels) because they hardly ever sting. Until you are sure about the harmlessness of these *angelitos* you remain suspicious. I confess that often during astronomic observations I almost dropped my instruments when I realized my face and hands were covered with these hairy bees. Our guides assured us that these bees only attacked when you annoyed them by picking them up by their legs. I did not try.

It was half past four in the afternoon when we finished our observations. Satisfied with the success of our journey we forgot that there might be dangers descending steep slopes covered with a smooth, slippery grass in the dark. We did not arrive at the valley bottom until ten at night. We were exhausted and thirsty after walking for fifteen hours, practically without stopping. The soles of our feet were cut and torn by the rough, rocky soil and the hard, dry grass stalks, for we had been forced to pull our boots off as the ground was too slippery. We spent the night at the foot of La Silla. Our friends at Caracas had been able to follow us on the summit with binoculars. They liked hearing our account of the expedition but were not happy with the result of our measurements, for La Silla was not as high as the highest mountains in the Pyrenees.

Earthquake in Caracas – Departure from Caracas – Gold mines – Sugar plantations

We left Caracas on the 7th of February, on a fresh afternoon, ready to begin our journey to the Orinoco. The memory of this period is today more painful than it was years ago. In those remote countries our friends have lost their lives in the bloody revolutions that gave them freedom and then alternatively deprived them of it. The house where we lived is now a heap of rubble. Terrible earthquakes have transformed the shape of the ground; the city I described has disappeared. On the same spot, on the fissured ground, another city is slowly being built. The ruins, tombs for a large population, have already turned into shelter for human beings.

I reckoned that it was my duty in this book to record all the data obtained from reliable sources concerning the seismic shocks that on the 26th of March 1812 destroyed the city of Caracas; in all the province of Venezuela more than 20,000 people perished. As a historian of nature, the traveller should note down the moment when great natural calamities happen, and investigate the causes and relations, and establish fixed points in the rapid course of time, in the transformations that succeed each other ceaselessly so that he can compare them with previous catastrophes.

On my arrival at Terra Firma I was struck by the correlation between two natural phenomena: the destruction of Cumaná on the 14th of December 1797 and volcanic eruptions in the smaller West Indian Islands. Something similar happened at Caracas on the 26th of March 1812. In 1797 the volcano on Guadeloupe Island, on the Cumaná coast, seemed to have reacted; fifteen years later another volcano on San Vincente also reacted, and its effects were felt as far as Caracas and the banks of the Apure. Probably both times the centre of the eruption was at an enormous depth in the earth, equidistant from the points on the earth's surface that felt the movement. The shock felt at Caracas in December 1811 was the only one that preceded the terrible catastrophe of the 26th of March 1812. In Caracas, and for 90 leagues around, not one drop of rain had fallen for five months up to the destruction of the capital. The 26th of March was a very hot day; there was no wind and no cloud. It was Ascension Day and most people had congregated in the churches. Nothing suggested the horrors to come. At seven minutes past four the first shock was felt. 'It was so violent that the church bells rang, and lasted five to six seconds. It was followed immediately by another lasting ten to twelve seconds when the ground seemed to ripple like boiling water. People thought the quake was over when an infernal din came from under the ground. It was like thunder but louder and longer than any tropical storm. Following this there was a vertical movement lasting three seconds followed by undulations. The shocks coming from these contrary movements tore the city apart. Thousands of people were trapped in the churches and houses.'

On the 8th of February we set off at sunrise to cross Higuerote, a group of tall mountains separating the valleys of Caracas and Aragua. Descending the woody slopes of Higuerote towards the south-west we reached the small village of San Pedro, 584 toises high, located in a basin where several valleys meet. Banana trees, potatoes and coffee grow there. In an inn (*pulpería*) we met several European Spaniards working at the Tobacco Office. Their bad temper contrasted with our mood. Tired by the route, they vented their anger by cursing the wretched country ('estas tierras infelices') where they were doomed to live, while we never wearied of admiring the wild scenery, the fertile earth and mild climate. From Las Lagunetas we descended into the Tuy river valley. This western slope is called Las Cocuyzas, and is covered with two plants with agave leaves; the maguey of Cocuzza and the maguey of Cocuy. The latter belongs to the *Yucca* genus. Its sweet fermented juice is distilled into an alcohol, and I have seen people eat its young green leaves. The fibres of the full-grown leaves are made into extremely long cords. At Caracas cathedral a maguey cord has suspended the weight of a 350-pound clock for fifteen years. We spent two very agreeable days at the plantation of Don José de Manterola who, when young, had been attached to the Spanish Legation in Russia. Brought up and protected by Sr de Xavedra, one of the more enlightened administrators in Caracas, de Manterola wanted to leave for Europe when that famous man became minister. The governor of the province, fearing de Manterola's prestige, arrested him in the harbour and when the order from Spain finally arrived to release him from such an unjust arrest the minister had fallen from grace.

The farm we lodged at was a fine sugar-cane plantation. The ground is smooth like the bed of a dried lake. The Tuy river winds through land covered with banana trees and a little wood of *Hura crepitans*, *Erythrina corallodendron*, and figs with nymphae leaves. The river is formed with quartz pebbles. I can think of no more pleasant bathe than that in the Tuy. The crystal-clear water remains at 18.6°c. This is cool for the climate; the sources of the river are in the surrounding mountains. The owner's house is situated on a hillock surrounded by huts for the negroes. Those who are married provide their own food. They are given, as everywhere in the Aragua valleys, a plot of land to cultivate, which they work on their Saturdays and Sundays, the free days of the week. They have chicken, and sometimes a pig. The owner boasts of their contentment in the same way that northern European landowners boast about the happy peasants on their land. The day we arrived three runaway negroes had been captured; newly bought slaves. I dreaded witnessing those punishments that ruin the charm of the countryside wherever there are slaves. Luckily, the blacks were treated humanely.

In this plantation, as in all the provinces of Venezuela, you can distinguish, from afar, three kinds of sugar cane by the colour of their leaves; the old Creole cane, Otaheite cane and Batavia cane. The first has a darker green leaf, a thinner stalk with knots close together. It was the first sugar cane introduced from India to Sicily, the Canaries and the West Indies. The second is lighter green; its stalk is fatter, more succulent. The whole plant seems more luxuriant. It arrived thanks to the voyages of Bougainville, Cook and Bligh. Bougainville brought it to Mauritius, where it went to Cayenne, Martinique and from 1792 to the rest of the West Indies. Otaheite sugar cane, the *to* of the islanders, is one of the most important agricultural acquisitions due to the voyages of naturalists. On the same plot of land it gives a third of *vezou* (juice) more than Creole cane, but due to the thickness of its stalk and strength of its ligneous fibres furnishes much more fuel. This is an advantage in the West Indian Islands where the destruction of the forests has forced planters to use the bagasse as fuel for their furnaces. The third species, the violet sugar cane, is called Batavia or Guinea cane, and certainly comes from Java. Its leaves are purple and large and it is preferred in Caracas for making rum. At Tuy they were busy finishing a ditch to bring irrigation water. This enterprise had cost the owner 7,000 piastres to build and 4,000 piastres in lawsuits with his neighbours. While the lawyers argued over the canal, which was only half finished, de Manterola had already begun to doubt the worth of his project. I took the level of the ground with a *lunette d'épreuve* placed on an artificial horizon and found that the dam had been placed 8 feet too low. What sums of money have not been uselessly spent in the Spanish colonies founding constructions on poor levelling!

The Tuy valley has its 'gold mine', as do nearly all the places near mountains inhabited by white Europeans. I was assured that in 1780 foreign gold seekers had been seen extracting gold nuggets and had set up a place for washing the sand. The overseer of a nearby plantation had followed their tracks and after his death a jacket with gold buttons was found among his belongings, which according to popular logic meant that they came from the gold seam, later covered by a rock fall. It was no use my saying that from simply looking at the ground, without opening up a deep gallery, I would not be able to decide if there once had been a mine there – I had to yield to my host's entreaties. For twenty years the overseer's jacket had been the talking-point of the area. Gold dug out from the ground has, in the people's eyes, a special lure unrelated to the diligent farmer harvesting a fertile land under a gentle climate.

Our guides led us to the 'mine'. We turned west, and finally reached the Quebrada de Oro. On the hillside there was hardly a trace of a quartz seam. The landslide, caused by rain, had so transformed the ground that we could not even think of exploring it. Huge trees now grew where twenty years before gold seekers had worked. It is likely that there are veins in the mica-slate containing this venerable metal, but how could I judge if it was worth exploiting or if the metal was to be found in nodules? To compensate our efforts, we set to botanizing in the thick wood around the Hato.

We left the Manterola plantation on the 11th of February at sunrise. A little before reaching Mamon we stopped at a farm belonging to the Monteras family. A negress, more than a hundred years old, was sitting outside a mud-and-reed hut. Her age was known because she had been a creole slave. She seemed to enjoy amazing good health. 'I keep her in the sun' (*La tengo al sol*), said her grandson. 'The heat keeps her alive.' This treatment seemed rather harsh as the sun's rays fell vertically on to her. Blacks and Indians reach very advanced ages in the torrid zone. Hilario Pari, a native of Peru, died at the extraordinary age of one hundred and forty-three, having been married ninety years.

Beyond the village of Turmero, towards Maracay, you can observe on the distant horizon something that seems to be a tumulus covered in vegetation. But it is not a hill, nor a group of trees growing close together, but one single tree, the famous *zamang de Guayre*, known through the country for the enormous extent of its branches, which form a semi-spherical head some 576 feet in circumference. The zamang is a fine species of the mimosa family whose twisted branches are forked. We rested a long time under this vegetable roof. The branches extend like an enormous umbrella and bend towards the ground. Parasitical plants grow on the branches and in the dried bark. The inhabitants, especially the Indians, venerate this tree, which the first conquerors found in more or less the same state as it is in today. We heard with satisfaction that the present owner of the zamang had brought a lawsuit against a cultivator accused of cutting off a branch. The case was tried and the man found guilty.

We reached Maracay late. The people who had been recommended to us were away, but no sooner had the inhabitants realized our worries than they came from everywhere to offer us lodging for our instruments and mules. It has been said a thousand times, but the traveller always feels the need to repeat that the Spanish colonies are the authentic land of hospitality, even in places where industry and commerce have created wealth and a little culture. A Canarian family warmly invited us to stay, and cooked an excellent dinner. The master of the house was away on a business trip and his young wife had just given birth. She was wild with joy when she heard that we were due to pass through Angostura where her husband was. Through us he would learn about the birth of his first child. As we were about to leave we were shown the baby; we had seen her the night before, asleep, but the mother wanted us to see her awake. We promised to describe her features one by one to the father, but when she saw our instruments and books the good woman worried: 'On such a long journey, and with so many other things to think about, you could easily forget the colour of my baby's eyes!'

We spent seven agreeable days at the Hacienda de Cura in a small hut surrounded by thickets; the house itself, located in a sugar plantation, was infected with bubos, a skin disease common among slaves in the valleys. We lived like the rich; we bathed twice a day, slept three times and ate three meals in twenty-four hours. The lake water was warm, some 24°C to 25°C. The coolest bathing place was under the shade of ceibas and zamangs at Toma in a stream that rushes out of the granite Rincón del Diablo mountains. Entering this bath was fearsome, not because of the insects but because of the little brown hairs covering the pods of the *Dolichos pruriens*. When these small hairs, called *pica pica*, stick to your body they cause violent irritations. You feel the sting but cannot see what stung you.

During our stay at Cura we made numerous excursions to the rocky islands in the middle of Lake Valencia, to the hot springs at Mariara, and the high mountain called El Cucurucho de Coco. A narrow, dangerous path leads to the port of Turiamo and the famous coastal cacao plantations. Throughout all our excursions we were surprised not only by the progress of culture but also by the increase in the numbers of the free, hard-working population, used to manual work and too poor to buy slaves. Everywhere whites and mulattos had bought small isolated farms. Our host, whose father enjoyed an income of 40,000 piastres a year, had more land than he could farm; he distributed plots in the Aragua valley to poor families who wanted to grow cotton. He tried to surround his enormous plantation with free working men, because they wanted to work for themselves, or for others. Count Tovar was busy trying to abolish slavery and hoped to make slaves less necessary for the important estates, and to offer the freed slaves land to become farmers themselves. When he left for Europe he had broken up and rented land around Cura. Four years later, on returning to America, he found fine cotton fields and a little village called Punta Samuro, which we often visited with him. The inhabitants are all mulatto, *zambo* [of mixed African and Indian parentage] and freed slaves. The rent is ten piastres a *fanega* of land; it is paid in cash or cotton. As the small farmers are often in need, they sell their cotton at modest prices. They sell it even before harvest, and this advance is used by the rich landowners to make the poor dependent on them as day workers. The price of labour is less than it is in France. A free man is paid five piastres a month without food, which costs very little as meat and vegetables are abundant. I like quoting these details about colonial agriculture because they prove to Europeans that there is no doubt that sugar, cotton and indigo can be produced by free men, and that the miserable slaves can become peasants, farmers and landowners.

Lake Tacarigua – Hot springs of Mariara – The town of Nueva Valencia – Descent to the Puerto Cabello coasts

The Aragua valleys form a basin, closed between granitic and calcareous mountain ranges of unequal height. Due to the land's peculiar configuration, the small rivers of the Aragua valleys form an enclosed system and flow into a basin blocked off on all sides; these rivers do not flow to the ocean but end in an inland lake, and thanks to constant evaporation lose themselves, so to speak, in the air. These rivers and lakes determine the fertility of the soil and agricultural produce in the valleys. The aspect of the place and the experience of some fifty years show that the water-level is not constant; that the balance between evaporation and inflow is broken. As the lake lies 1,000 feet above the neighbouring Calabozo steppes, and 1,332 feet above sea-level, it was thought that the water filtered out through a subterranean channel. As islands emerge, and the water-level progressively decreases, it is feared the lake might completely dry out.

Lake Valencia, called Tacarigua by the Indians, is larger than Lake Neuchâtel in Switzerland; its general form resembles Lake Geneva, situated at about the same altitude. Its opposite banks are notably different: the southern one is deserted, stripped of vegetation and virtually uninhabited; a curtain of high mountains gives it a sad, monotonous quality; in contrast, the northern side is pleasant and rural, and has rich plantations of sugar cane, coffee and cotton. Paths bordered with cestrum, azedaracs, and other perpetually flowering shrubs cross the plain and link the isolated farms. All the houses are surrounded by trees. The ceiba (*Bombax hibiscifolius*), with large yellow flowers, and the erythrina, with purple ones, whose overlapping branches give the countryside its special quality. During the season of drought, when a thick mist floats above the burning ground, artificial irrigation keeps the land green and wild. Every now and then granite blocks pierce through the cultivated ground; large masses of rocks rise up in the middle of the valley. Some succulent plants grow in its bare and cracked walls, preparing mould for the coming centuries. Often a fig tree, or a clusia with fleshy leaves, growing in clefts, crowns these isolated little summits. With their dry withered branches they look like signals along a cliff. The shape of these heights betrays the secret of their ancient origins; for when the whole valley was still submerged and waves lapped the foot of the Mariara peaks (El Rincón del Diablo) and the coastal chain, these rocky hills were shoals and islands.

But the shores of Lake Valencia are not famed solely for their picturesque beauties: the basin presents several phenomena whose interpretation holds great interest for natural historians and for the inhabitants. What causes the lowering of the lake's water-level? Is it receding faster than before? Will the balance between the flowing in and the draining out be restored, or will the fear that the lake might dry up be proved justified?

I have no doubt that from remotest times the whole valley was filled with water. Everywhere the shape of the promontories and their steep slopes reveals the ancient shore of this alpine lake. We find vast tracts of land, formerly flooded, now cultivated with banana, sugar cane and cotton. Wherever a hut is built on the lake shore you can see how year by year the water recedes. As the water decreases, you can see how islands begin to join the land while others form promontories or become hills. We visited two islands still completely surrounded by water and found, under the scrub, on small flats between 4 and 8 toises above the water-level, fine sand mixed with helicites deposited by waves. On all these islands you will discover clear traces of the gradual lowering of the water.

The destruction of the forests, the clearing of the plains, and the cultivation of indigo over half a century has affected the amount of water flowing in as well as the evaporation of the soil and the dryness of the air, which forcefully explains why the present Lake Valencia is decreasing. By felling trees that cover the tops and sides of mountains men everywhere have ensured two calamities at the same time for the future: lack of fuel, and scarcity of water. Trees, by the nature of their perspiration, and the radiation from their leaves in a cloudless sky, surround themselves with an atmosphere that is constantly cool and misty. They affect the amount of springs by sheltering the soil from the sun's direct actions and reducing the rainwater's evaporation.

When forests are destroyed, as they are everywhere in America by European planters, with imprudent haste, the springs dry up completely, or merely trickle. River beds remain dry part of the year and are then turned into torrents whenever it rains heavily on the heights. As grass and moss disappear with the brushwood from the mountainsides, so rainwater is unchecked in its course. Instead of slowly raising the river level by filtrations, the heavy rains dig channels into the hillsides, dragging down loose soil, and forming sudden, destructive floods. Thus, the clearing of forests, the absence of permanent springs, and torrents are three closely connected phenomena. Countries in different hemispheres like Lombardy bordered by the Alps, and Lower Peru between the Pacific and the Andes, confirm this assertion.

Until the middle of the last century the mountains surrounding the Aragua valley were covered in forests. Huge trees of the mimosa, ceiba and fig families shaded the lake shore and kept it cool. The sparsely populated plain was invaded by shrubs, fallen tree trunks and parasitical plants, and was covered in thick grass so that heat was not lost as easily as from cultivated ground, which is not sheltered from the sun's rays. When the trees are felled, and sugar cane, indigo and cotton are planted, springs and natural supplies to the lake dry up. It is hard to form a fair idea of the enormous amount of evaporation taking place in the torrid zone, especially in a valley surrounded by steep mountains where maritime breezes blow, and whose ground is completely flat as if levelled by water. The heat prevailing on the lake shore is comparable to that in Naples and Sicily.

Lake Valencia is full of islands, which embellish the countryside with the picturesque form of their rocks and by the kind of vegetation that covers them. Tropical lakes have this advantage over alpine ones. The islands, without counting Morro and Cabrera, which are already joined to the mainland, are fifteen in number. They are partially cultivated, and very fertile due to the vapours rising from the lake. Burro, the largest island, some 2 miles long, is inhabited by mestizo families who rear goats. These simple people rarely visit the Mocundo coast. The lake seems gigantic to them: they produce bananas, cassava, milk and fish. A hut built of reeds, some hammocks woven with cotton grown in neighbouring fields, a large stone on which they build their fires, and the ligneous fruit of the *tutuma* to draw water with are their sole household needs. The old mestizo who offered us goat's milk had a lovely daughter. We learned from our guide that isolation had made him as suspicious as if he lived in a city. The night before our arrival some hunters had visited the island. Night surprised them and they preferred to sleep out in the open rather than return to Mocundo. This news spread alarm around the island. The father forced his young daughter to climb a very tall zamang or mimosa, which grows on the plain at some distance from the hut. He slept at the foot of this tree, and didn't let his daughter down until the hunters had left.

The lake is usually full of fish; there are three species

with soft flesh, which are not very tasty: the *guavina*, the *bagre* and the *sardina*. The last two reach the lake from streams. The *guavina*, which I sketched on the spot, was some 20 inches long and 3 to 5 inches wide. It is perhaps a new species of Gronovius's *Erythrina*. It has silver scales bordered with green. This fish is extremely voracious and destroys other species. Fishermen assured us that a little crocodile, the *bava*, which often swam near as we bathed, contributed to the destruction of the fish. We never managed to catch this reptile and examine it close up. It is said to be very innocent; yet its habits, like its shape, clearly resemble the alligator or *Crocodilus acutus*. It swims so that only the tips of its snout and tail show: it lies at midday on deserted beaches.

The island of Chamberg is a granitic outcrop some 200 feet high, with two peaks linked by a saddle. The sides of the rock are bare; only a few white flowering clusia manage to grow there. But the view of the lake and surrounding plantations is magnificent, especially at sunset when thousands of heron, flamingo and wild duck fly over the water to roost on the island.

It is thought that some of the plants that grow on the rocky islands of Lake Valencia are exclusive to them because they have not been discovered elsewhere. Among these are the papaw tree of the lake (*papaya de la laguna*), and a tomato from Cura Island; this differs from our *Solanum lycopersicum* in that its fruit is round and small but very tasty. The papaw of the lake is common also on Cura Island and at Cabo Blanco. Its trunk is slenderer than the ordinary papaw, but its fruit is half the size and completely round, without projecting ribs. This fruit, which I have often eaten, is extremely sweet.

The areas around the lake are unhealthy only in the dry season when the water-level falls and the mud bed is exposed to the sun's heat. The bank, shaded by woods of *Coccoloba barbadensis* and decorated with beautiful lilies, reminds one, because of the similar aquatic plants found there, of the marshy banks of our European lakes. Here we find pondweed (*potamogeton*), chara and cat's-tails 3 feet high, hardly different from the *Typha angustifolia* of our marshes. Only after very careful examination do we recognize each plant to be a distinct species, peculiar to the New World. How many plants from the Strait of Magellan to the cordilleras of Quito have once been confused with northern temperate ones owing to their analogy in form and appearance!

Some of the rivers flowing into Lake Valencia come from thermal springs, worthy of special note. These springs gush out at three points from the coastal granitic chain at Onoto, Mariara and Las Trincheras. I was only able to carefully examine the physical and geological relations of the thermal waters of Mariara and Las Trincheras. All the springs contain small amounts of sulphuretted hydrogen gas. The stink of rotten eggs, typical of this gas, could only be smelled very close to the spring. In one of the puddles, which had a temperature of 56.2°c, bubbles burst up at regular intervals of two to three minutes. I was not able to ignite the gas, not even the small amounts in the bubbles as they burst on the warm surface of the water, nor after collecting it in a bottle, despite feeling nausea caused more by the heat than by the gas. The water, when cold, is tasteless and quite drinkable.

South of the ravine, in the plain that stretches to the lake shore, another less hot and less gassy sulphureous spring gushes out. The thermometer reached only 42°c. The water collects in a basin surrounded by large trees. The unhappy slaves throw themselves in this pool at sunset, covered in dust after working in the indigo and sugar-cane fields. Despite the water being 12°c to 14°c warmer than the air the negroes call it refreshing. In the torrid zone this word is used for anything that restores your strength, calms nerves or produces a feeling of well-being. We also experienced the salutary effects of this bath. We had our hammocks slung in the trees shading this pond and spent a whole day in this place so rich in plants. Near this *bāno de Mariara* we found the *volador* or gyrocarpus. The winged fruits of this tree seem like flying beings when they separate from the stem. On shaking the branches of the *volador*, we saw the air filled with its fruits, all falling together. We sent some fruit to Europe, and they germinated in Berlin, Paris and Malmaison. The numerous plants of the *volador*, now seen in hothouses, owe their origin to the only tree of its kind found near Mariara.

While following the local custom of drying ourselves in the sun after our bath, half wrapped in towels, a small mulatto approached. After greeting us in a serious manner, he made a long speech about the properties of the Mariara waters, the many sick people who over the years have come here, and the advantageous position of the spring between Valencia and Caracas, where morals became more and more dissolute. He showed us his house, a little hut covered with palm leaves in an enclosure near by, next to a stream that fed the pool. He assured us that we would find there all the comforts we could imagine; nails to hang our hammocks, oxhides to cover reed beds, jugs of fresh water, and those large lizards (iguanas) whose flesh is considered to be a refreshing meal after a bathe. From his speech we reckoned that this poor man had mistaken us for sick people wanting to install themselves near the spring. He called himself 'the inspector of the waters and the *pulpero* of the place'. He stopped talking to us as soon as he saw we were there out of curiosity – 'para ver no más' as they say in these colonies, 'an ideal place for lazy people'.

On the 21st of February, at nightfall, we left the pretty Hacienda de Cura and set off for Guacara and Nueva Valencia. As the heat of the day was stifling we travelled by night. We crossed the village of Punta Zamuro at the foot of Las Viruelas mountain. The road is lined with large zamangs, or mimosa trees, reaching some 60 feet high. Their almost horizontal branches meet at more than 150 feet distance. I have never seen a canopy of leaves so thick and beautiful as these. The night was dark: the Rincón del Diablo and its dentated rocks appeared every now and then, illuminated by the brilliance of the burning savannahs, or wrapped in clouds of reddish smoke. In the thickest part of the brush our horses panicked when they heard the howl of an animal that seemed to be following us. It was an enormous jaguar that had been roaming these mountains for three years. It had escaped from the most daring hunters. It attacked horses and mules, even when they were penned in, but not lacking food had not yet attacked human beings. Our negro guide screamed wildly to scare off the beast, which he obviously did not achieve.

We spent the 23rd of February in the marquis of Toro's house, in the village of Guacara, a large Indian community. The Indians live a life of ease because they have just won a legal case restoring lands disputed by whites. An avenue of carolineas leads from Guacara to Mocundo, a rich sugar plantation belonging to the Moro family. We found a rare garden there with an artificial clump of trees, and, on top of a granitic outcrop near a stream, a pavilion with a *mirador* or viewpoint. From here you see a splendid panorama over the west of the lake, the surrounding mountains and a wood of palm trees. The sugar-cane fields with their tender green leaves seem like a great plain. Everything suggests abundance, although those who work the land have to sacrifice their freedom.

The preparation of sugar, its boiling, and the claying, is not well done in Terra Firma because it is made for local consumption. More *papelón* is sold than either refined or raw sugar. *Papelón* is an impure sugar in the form of little yellowish-brown loaves. It is a blend of molasses and mucilaginous matter. The poorest man eats *papelón* the way in Europe he eats cheese. It is said to be nutritious. Fermented with water it yields *guarapo*, the favourite local drink.

The city of New Valencia occupies a large area of ground, but its population is of some 6,000 to 7,000 souls. The roads are very wide, the market place (*plaza mayor*) is disproportionately large. As the houses are few the difference between the population and the land they occupy is greater even than at Caracas. Many of the whites of European stock, especially the poorest, leave their town houses and live for most of the year in their cotton and indigo plantations. They dare to work with their own hands, which, given the rigid prejudices in this country, would be a disgrace in the city. The industriousness of the inhabitants has greatly increased after freedom was granted to business in Puerto Cabello, now open as a major port (*puerto mayor*) to ships coming directly from Spain.

Founded in 1555, under the government of Villacinda, by Alonso Díaz Moreno, Nueva Valencia is twelve years older than Caracas. Some justifiably regret that Valencia has not become the capital of the country. Its situation on the plain, next to a lake, recalls Mexico City. If you consider the easy communications offered by the Aragua valleys with the plains and rivers entering the Orinoco; if you accept the possibility of opening up navigation into the interior through the Pao and Portuguesa rivers as far as the Orinoco mouth, the Casiquiare and the Amazon, you realize that the capital of the vast Venezuelan provinces would have been better placed next to the superb Puerto Cabello, under a pure, serene sky, and not next to the barely sheltered bay of La Guaira, in a temperate but always misty valley.

Only those who have seen the quantity of ants that infest the countries of the torrid zone can picture the destruction and the sinking of the ground caused by these insects. They abound to such a degree in Valencia that their excavations resemble underground canals, which flood with water during the rains and threaten buildings. Here they have not used the extraordinary means employed by the monks on the island of Santo Domingo when troops of ants ravaged the fine plains of La Vega. The monks, after trying to burn the ant larvae and fumigate the nests, told the inhabitants to choose a saint by lot who would act as an Abogado contra las Hormigas. The choice fell on Saint Saturnin, and the ants disappeared as soon as the saint's festival was celebrated.

On the morning of the 27th of February we visited the hot springs of La Trinchera, 3 leagues from Valencia. They flow more fully than any we had seen until then, forming a rivulet, which in the dry season maintains a depth of some 2 feet 8 inches of water. The carefully taken water temperature was 90.3°C. We had breakfast near the spring: our eggs were cooked in less than four minutes in the hot water. The rock from which the spring gushes is of real coarse-grained granite. Whenever the water evaporates in the air, it forms sediments and incrustations of carbonate of lime. The exuberance of the vegetation around the basin surprised us. Mimosas with delicate pinnate leaves, clusias and figs send their roots into the muddy ground, which is as hot as 85°c. Two currents flow down on parallel courses, and the Indians showed us how to prepare a bath of whatever temperature you want by opening a hole in the ground between the two streams. The sick, who come to La Trinchera to take steam baths, build a kind of framework with branches and thin reeds above the spring. They lie down naked on this frame, which, as far as I could see, was not very strong, perhaps even dangerous.

As we approached the coast the heat became stifling. A reddish mist covered the horizon. It was sunset but no sea breeze blew. We rested in the lonely farm called both Cambury and House of the Canarian (Casa del Isleño). The hot-water river, along whose bank we travelled, became deeper. A 9-foot-long crocodile lay dead on the sand. We wanted to examine its teeth and the inside of its mouth, but having been exposed to the sun for weeks it stank so badly we had to climb back on to our horses.

More than 10,000 mules are exported every year from Puerto Cabello. It is curious to see these animals being embarked. They are pulled down with lassos and lifted on board by something akin to a crane. In the boat they are placed in double rows, and with the rolling and pitching of the boat can barely stand. To terrify them, and keep them docile, a drum is beaten day and night.

From Puerto Cabello we returned to the Aragua valley, and stopped again at the Barbula plantation through which the new road to Nueva Valencia will pass. Weeks before we had been told about a tree whose sap is a nourishing milk. They call it the 'cow tree', and assured us that negroes on the estates drank quantities of this vegetable milk. As the milky juices of plants are acrid, bitter and more or less poisonous, it seemed hard to believe what we heard, but during our stay in Barbula we proved that nobody had exaggerated the properties of *palo de vaca*. This fine tree is similar to the *Chrysophyllum cainito* (broad-leafed star-apple). When incisions are made in the trunk it yields abundant glutinous milk; it is quite thick, devoid of all acridity, and has an agreeable balmy smell. It was offered to us in *tutuma*-fruit – or ground – bowls, and we drank a lot before going to bed, and again in the morning, without any ill effects. Only its viscosity makes it a little disagreeable. Negroes and free people who work on the plantations dip their maize and cassava bread in it. The overseer of the estate told us that negroes put on weight during the period that the *palo de vaca* exudes milk. This notable tree appears to be peculiar to the cordillera coast. At Caucagua the natives called it the 'milk tree'. They say they can recognize the trunks that yield most juice from the thickness and colour of the leaves. No botanist has so far known this plant.

Of all the natural phenomena that I have seen during my voyages few have produced a greater impression than the *palo de vaca*. What moved me so deeply was not the proud shadows of the jungles, nor the majestic flow of the rivers, nor the mountains covered with eternal snows, but a few drops of a vegetable juice that brings to mind all the power and fertility of nature. On a barren rocky wall grows a tree with dry leathery leaves; its large woody roots hardly dig into the rocky ground. For months not a drop of rain wets its leaves; the branches appear dry, dead. But if you perforate the trunk, especially at dawn, a sweet nutritious milk pours out.

It was Carnival Tuesday, and everywhere people celebrated. The amusements, called *carnes tollendas* (or 'farewell to the flesh'), became at times rather wild: some paraded an ass loaded with water, and whenever they found an open window pumped water into the room; others carried bags full of hair from the *pica pica* (*Dolichos pruriens*), which greatly irritates skin on contact, and threw it into the faces of passers-by.

From La Guaira we returned to Nueva Valencia, where we met several French *émigrés*, the only ones we saw in five years in the Spanish colonies. In spite of the blood links between the Spanish and French royal families, not even French priests could find refuge in this part of the New World, where man finds it so easy to find food and shelter. Beyond the Atlantic Ocean, only the United States of America offers asylum to those in need. A government that is strong because it is free, and confident because it is just, has nothing to fear in granting refuge to exiles.

Before leaving the Aragua valleys and its neighbouring coasts, I will deal with the cacao plantations, which have always been the main source of wealth in this area. The cacao-producing tree does not grow wild anywhere in the forests north of the Orinoco. This scarcity of wild cacao trees in South America is a curious phenomenon, yet little studied. The amount of trees in the cacao plantations has been estimated at more than 16 million. We met no tribe on the Orinoco that prepared a drink with cacao seeds. Indians suck the pulp of the pod and chuck the seeds, often found in heaps in places where Indians have spent the night. It seems to me that in Caracas cacao cultivation follows the examples of Mexico and Guatemala. Spaniards established in Terra Firma learned how to cultivate the cacao tree – sheltered while young by the leaves of the erythrina and banana, making *chocolatl* cakes, and using the liquid of the same name, thanks to trade with Mexico, Guatemala and Nicaragua whose people are of Toltec and Aztec origin.

As far back as the sixteenth century travellers have greatly differed in their opinions about *chocolatl*. Benzoni said, in his crude language, that it is a drink 'fitter for pigs than humans'. The Jesuit Acosta asserts that 'the Spaniards who inhabit America are fond of chocolate to excess . . .' Fernando Cortez highly praised chocolate as being an agreeable drink if prepared cold and, especially, as being very nutritious. Cortez's page writes: 'He who has drunk one cup can travel all day without further food, especially in very hot climates.' We shall soon celebrate this quality in chocolate in our voyage up the Orinoco. It is easily transported and prepared: as food it is both nutritious and stimulating.

Mountains situated between the Aragua valleys and the Caracas plains – Villa de Cura – Parapara – Llanos or steppes – Calabozo

The chain of mountains limited on the south by Lake Tacarigua forms, you could say, the northern boundary of the great basin of the plains or savannahs of Caracas. From the Aragua valleys you reach the savannahs over the Guigue and Tucutenemo mountains. Moving from a region peopled and embellished by agriculture you find a vast desert. Accustomed to rocks and shaded valleys, the traveller contemplates with astonishment those plains without trees, those immense tracts of land that seem to climb to the horizon.

We left the Aragua valleys before sunset on the 6th of March. We crossed a richly cultivated plain, bordering the south-westerly banks of Lake Valencia, along ground recently uncovered by receding water. The fertility of the earth, planted with gourds, water melons and bananas, amazed us. The distant howling of monkeys announced dawn. Opposite a clump of trees in the middle of the plain we caught sight of several bands of araguatoes (*Simia ursina*) who, as if in procession, passed very slowly from one branch to another. After the male followed several females, many with young on their backs. Due to their life-style howling monkeys all look alike, even those belonging to

45

different species. It is striking how uniform their movements are. When the branches of two trees are too far apart, the male that guides his troop hangs on his prehensile tail and swings in the air until he reaches the nearest branch. Then all the band repeat the same operation in the same place. It is almost superfluous to add how dubious Ulloa's assertion is that the araguatoes form a kind of chain in order to reach the opposite bank of a river. During five years we had ample opportunity to observe thousands of these animals: for this reason we have no confidence in statements possibly invented by Europeans themselves, although missionary Indians repeat them as if they come from their own traditions. The further man is from civilization, the more he enjoys astonishing people while recounting the marvels of his country. He says he has seen what he imagines may have been seen by others. Every Indian is a hunter and the stories of hunters borrow from the imagination the more intelligent the hunted animal appears to be. Hence so many fictions in Europe about the foxes, monkeys, crows and condors in the Andes.

The Indians claim that when howler monkeys fill the jungles with their howls there is always one that leads the howling. Their observation is correct. You generally hear one solitary and intense voice, replaced by another at a different pitch. Indians also assert that when an araguato female is about to give birth, the chorus of howling stops until the new monkey is born. I was not able to prove this, but I have observed that the howling ceases for a few minutes when something unexpected happens, like when a wounded monkey claims the attention of the troop. Our guides seriously assured us that 'To cure asthma you must drink out of the bony drum of the araguato's hyoid bone.' Having such a loud voice this animal is thought to impart a curing effect from its larynx to the water drunk out of it. Such is the people's science, which sometimes resembles the ancients'.

We spent the night in the village of Guigue. We lodged with an old sergeant from Murcia. To prove he had studied with the Jesuits he recited to us the history of the creation in Latin. He knew the names of Augustus, Tiberius and Diocletian, and while enjoying the agreeably cool nights on his banana plantation interested himself in all that had happened in the times of the Roman emperors. He asked us for a remedy for his painful gout. 'I know,' he said, 'that a *zambo* from Valencia, a famous *curioso*, could cure me, but the *zambo* would expect to be treated as an equal, and that I cannot do with a man of his colour. I prefer to remain as I am.'

San Luis de Cura or, as it is more usually called, Villa de Cura, lies in a very barren valley. Apart from a few fruit trees the region is without vegetation. The *meseta* is dry and several rivers lose themselves in cracks in the ground. Cura is more a village than a town. We lodged with a family that had been persecuted by the government after the 1797 revolution in Caracas. After years in prison, one of their sons had been taken to Havana, where he lived locked in a fort. How pleased his mother was when she heard that we were bound for Havana after visiting the Orinoco. She handed me

five piastres – 'all her savings'. I tried to hand them
back, but how could I wound the delicacy of a woman
happy with her self-imposed sacrifice! All the society
in the village met in the evening to look at a magic
lantern showing sights of the great European cities,
the Tuileries palace and the statue of the Great Elector
in Berlin. How odd to see our native city in a magic
lantern some 2,000 leagues away!

After bathing in the fresh clear water of the San
Juan river at two in the morning, we set off on the road
for Mesa de Paja. The llanos at that time were infested
with bandits, so other travellers joined us to form a
kind of caravan. The route was downhill for several
hours.

At Mesa de Paja we entered the basin of the llanos.
The sun was almost at its highest point. On the ground
we recorded a temperature of 48°c to 50°c in the
sterile parts without vegetation. At the height of our
heads, as we were riding the mules, we did not feel the
slightest breath of air; but in the midst of that apparent
calm small dust whirls were continually raised by air
currents arising from the difference in temperature
between the bare sand and the grass. These sand winds
increased the suffocating heat. The plains surrounding
us seemed to reach the sky and looked to us like an
ocean covered with seaweed. Sky and land merged.
Through the dry mist and vapours you could make out,
in the distance, trunks of palm trees. Stripped of their
leaves these trunks looked like ship masts on the
horizon.

The monotony of these steppes is imposing, sad and

oppressive. Everything appears motionless; only now and then from a distance does the shadow of a small cloud promising rain move across the sky. The first glimpse of the plains is no less surprising than that of the Andean chain. It is hard to get accustomed to the views on the Venezuelan and Casanare plains, or to the pampas of Buenos Aires and the *chaco* when, for twenty to thirty days without stopping, you feel you are on the surface of an ocean. The plains of eastern and northern Europe can give only a pallid image of the immense South American llanos.

The llanos and pampas of South America are really steppes. During the rainy season they appear beautifully green, but in the dry season they look more like deserts. The grass dries out and turns to dust; the ground cracks, crocodiles and snakes bury themselves in the dried mud waiting for the first rains of spring to wake them from prolonged lethargy.

Rivers have only a slight, often imperceptible fall. When the wind blows, or the Orinoco floods, the rivers disemboguing in it are pushed backward. In the Arauca you often see the current going the wrong way. Indians have paddled a whole day downstream when in reality they have been going upstream. Between the descending and ascending waters lie large stagnant tracts, and dangerous whirlpools are formed.

The most typical characteristic of the South American savannahs or steppes is the total absence of hills, the perfect flatness of the land. That is why the Spanish conquistadores did not call them deserts, savannahs or meadows but plains, *los llanos*. Often in an area of

600 square kilometres no part of the ground rises more than 1 metre high.

Despite the apparent uniformity of the ground the llanos offer two kinds of inequalities that cannot escape the attentive traveller. The first are called *bancos* (banks); they are in reality shoals in the basin of the steppes, rising some 4 to 5 feet above the plains. These banks can reach some 3 to 4 leagues in length; they are completely smooth and horizontal, and can only be recognized when you examine their edges. The second inequality can only be detected by geodesical or barometric measurements, or else by the flow of a river; they are called *mesa*, or tables. They are small flats, or convex elevations, that rise imperceptibly some metres high to divide the waters between the Orinoco and the northern Terra Firma coast. Only the gentle curvature of the savannah forms this division.

The infinite monotony of the llanos; the extreme rarity of inhabitants; the difficulties of travelling in such heat and in an atmosphere darkened by dust; the perspective of the horizon, which constantly retreats before the traveller; the few scattered palms that are so similar that one despairs of ever reaching them, and confuses them with others further afield; all these aspects together make the stranger looking at the llanos think they are far larger than they are.

After spending two nights on horseback, and having vainly looked for shade under tufts of the mauritia palms, we arrived before nightfall at the small farm called Alligator (El Cayman), also called La Guadalupe. It is an *hato de ganado*, that is, an isolated house on the steppes, surrounded by small huts covered in reeds and skins. Cattle, oxen, horses and mules are not penned in; they wander freely in a space of several square leagues. Nowhere do you see any enclosures. Men, naked to the waist, and armed with lances, ride the savannahs to inspect the animals, to bring back those that have strayed too far off, and to brand with a hot iron those still not branded with the owner's mark. These coloured men, called *peones llaneros*, are partly freed and partly enslaved. There is no race more constantly exposed to the devouring fire of the tropical sun than this one. They eat meat dried in the sun, and barely salted. Even their horses eat this. Always in the saddle, they do not ever try to walk a few paces. On the farm we found an old negro slave in charge while his master was away. We were told about herds of several thousand cows grazing the steppes, and yet it was impossible to get a bowl of milk. We were offered a yellowish, muddy and fetid water drawn from a nearby stagnant pool in bowls made of *tutuma* fruit. The laziness of the llano inhabitants is such that they cannot be bothered to dig wells, even though they know that 9 feet down you can everywhere find fine springs under a stratum of conglomerate or red sandstone. After suffering half a year of flooding, you are then exposed to another half of painful drought. The old negro warned us to cover the jug with a cloth and to drink the water through a filter so as not to smell the stink, and not to swallow the fine yellowish clay in the water. We did not know then that we would follow his instructions for months on end. The Orinoco waters are just as charged with particles of earth, and are even fetid in creeks, where dead crocodiles rot on sandbanks, half buried in the slime.

We had hardly unpacked our instruments before we freed our mules and let them, as is said here, 'find water on the savannah'. There are small pools around the farm and animals find them guided by instinct, by the sight of scattered tufts of mauritia palms, by the sensation of humidity that gives rise to small air currents in an otherwise calm atmosphere. When these stagnant ponds are far off, and the farm-hands are too lazy to lead the animals to their natural watering-holes, they are locked for five or six hours in a very hot stable, and then released. Excessive thirst increases their instinctive cleverness. As soon as you open the stable doors you see the horses, and especially the mules, far more intelligent than horses, rush off into the savannah. Tails in the air, heads back, they rush into the wind, stopping for a while to explore around them, following less their sight than their sense of smell, until they finally announce by neighing that water has been found. All these movements are more successfully carried out by horses born on the llanos who have enjoyed the freedom of wild herds than by those coming from the coast, descendants of domestic horses. With most animals, as with man, the alertness of the senses diminishes after years of work, after domestic habits and the progress of culture.

We followed our mules as they sought one of these stagnant ponds that give muddy water, which hardly satisfied our thirst. We were covered in dust, and tanned by the sand wind, which burns the skin more than the sun. We were desperate to have a bathe but we found only a pool of stagnating water surrounded by palms. The water was muddy, but to our surprise cooler than the air. Used as we were on this long journey to bathing every time we could, often several times a day, we did not hesitate to throw ourselves into the pool. We had hardly begun to enjoy the cool water when we heard a noise on the far bank that made us leap out. It was a crocodile slipping into the mud. It would have been unwise to spend the night in that muddy place.

We had gone scarcely more than a quarter of a league away from the farm, yet we walked for more than an hour on our way back without reaching it. Too late we saw that we had been going in the wrong direction. We had left as the day ended, before the stars had come out, and had proceeded haphazardly into the plains. As usual we had our compass. It would have been easy to find our direction from the position of Canopus and the Southern Cross; but the means were useless because we were uncertain whether we had gone east or south when we left the small farm. We tried to return to our bathing place, and walked for another three quarters of an hour without finding the stagnant pond. We often thought we saw fire on the horizon; it was a star rising, its image magnified by vapours. After wandering for a long time on the savannah we decided to sit down on a palm trunk in a dry place surrounded by short grass; for Europeans who have recently arrived fear water snakes more than they

53

do jaguars. We did not fool ourselves into believing that our guides, whose indolence we well knew, would come looking for us before preparing and eating their food. The more unsure we were about our situation, the more pleasing it was eventually to hear horse hooves approaching from afar. It was an Indian, with his lance, doing his *rodeo*, that is, rounding up cattle. The sight of two white men saying they were lost made him think it was a trick. It was hard to convince him of our sincerity. He eventually agreed to lead us to the Alligator farm, but without slowing down his trotting horse. Our guides assured us that 'they were already getting worried about us', and to justify their worry had made a long list of people who had been lost in the llanos and found completely worn out. It is clear that danger exists only for those far from any farm or, as had happened recently, for those robbed by bandits and tied to a palm tree.

To avoid suffering the heat of day we left at two in the morning, hoping to reach Calabozo, a busy little town in the middle of the llanos, by midday. The appearance of the countryside remained always the same. There was no moon, but the great mass of stars decorating the southern skies lit up part of our path. This imposing spectacle of the starry vault stretching out over our heads, this fresh breeze blowing over the plains at night, the rippling of the grass wherever it is long, all reminded us of the surface of an ocean. This illusion increased especially (and we did not tire of the repetition of this sight) when the sun's disc showed on the horizon, doubling itself through refraction, and soon losing its flattened form, rising quickly towards the zenith.

As the sun rose the plains came alive. Cattle, lying down at night by ponds or at the foot of moriche and rhopala palms, regrouped, and the solitudes became populated with horses, mules and oxen that live here not like wild animals but free, without fixed abode, scorning man's care. In this torrid zone the bulls, although of Spanish pedigree like those on the cold tablelands of Quito, are tame. The traveller is never in danger of being attacked or chased, contrary to what often happened during our wanderings in the Andes. Near Calabozo we saw herds of roebucks grazing peacefully with the horses and oxen. They are called *matacanes*; their meat is very tasty. They are larger than our deer and have a very sleek skin of a dark brown with white spots. Their horns seem to be simple points and they are not shy. We saw some completely white ones in the groups of thirty to forty that we observed.

Besides the scattered trunk of the *palma de cobija* we found real groves (*palmares*) in which the corypha is mixed with a tree of the proteaceous family called *chaparro* by the Indians, which is a new species of rhopala, with hard, crackling leaves. The little groves of rhopala are called *chaparrales* and it is easy to see that in a vast plain where only two or three kinds of tree grow that the *chaparro*, which gives shade, is deemed of great value. South of Guayaval other palms predominate: the *piritu* (*Bactris speciosa*) and the *mauritia* (*Mauritia flexuosa*), celebrated as the *árbol de la vida*. This last is the sago tree of America: it gives flour, wine, fibres to weave hammocks, baskets, nets and clothes. Its fruit, shaped like a pine-cone and covered in scales, tastes rather like an apple, and when ripe is yellow inside and red outside. Howler monkeys love them, and the Guaramo Indians, whose existence is closely linked to this palm, make a fermented liquor that is acid and refreshing.

On the La Mesa road, near Calabozo, it was extremely hot. The temperature of the air rose considerably as soon as the wind blew. The air was full of dust, and when there were gusts the thermometer reached 40°C and 41°C. We moved forward slowly as it would have been dangerous to leave the mules transporting our instruments behind. Our guides advised us to line our hats with rhopala leaves to mitigate the effect of the sun's rays on our heads. In fact it was quite a relief, and later we bore this in mind.

It is hard to formulate exactly how many cattle there are on the llanos of Caracas, Barcelona, Cumaná and Spanish Guiana. Monsieur Depons, who has lived longer in Caracas than I have, and whose statistics are generally correct, calculates that in these vast plains, from the mouth of the Orinoco to Lake Maracaibo, there are 1,200,000 oxen, 180,000 horses and 90,000 mules. He worked out a value of 5 million francs for the produce of these herds, including exportation and the price of leather in the country. In the Buenos Aires pampas there are, so we believe, some 12 million cows and 3 million horses, not counting the animals without owners.

I shall not hazard any general evaluations as they are

too vague by nature; but I will observe that in the Caracas llanos owners of the great *hatos* have no idea how many animals they have. They count only the young animals branded every year with the sign of their herd. The richer owners brand up to 14,000 animals a year, and sell 5,000 to 6,000. According to official documents the export of leather in all the Capitanía-General of Caracas reaches 174,000 oxhides and 11,500 goat hides. When one remembers that these figures come from custom registers and do not include contra-band one is tempted to think that the calculation of 1,200,000 oxen wandering in the llanos is far too low.

In Calabozo, in the middle of the llanos, we found an electric machine with great discs, electrophori, batteries and electrometers; an apparatus as complete as any found in Europe. These objects had not been bought in America but made by a man who had never seen any instruments, who had never been able to consult anybody, and who knew about electricity only from reading Sigaud de la Fond's *Traité* and Franklin's *Mémoires*. Carlos del Pozo, this man's name, had begun by making cylindrical electrical machines using large glass jars, and cutting off their necks. Years later he managed to get two plates from Philadelphia to make a disc machine to obtain greater electric effects. It is easy to guess how difficult it must have been for Sr Pozo to succeed once the first works on electricity fell into his hands, and how he managed to work every-thing out for himself. Up to then he had enjoyed astonishing uneducated people with his experiments, and had never travelled out of the llanos. Our stay in

Calabozo gave him altogether another kind of pleasure. He must have set some value on two travellers who could compare his apparatus with European ones. With me I had electrometers mounted in straw, pith-balls and gold leaf, as well as a small Leyden jar that could be charged by rubbing, following Ingenhousz's method, which I used for physiological tests. Pozo could not hide his joy when for the first time he saw instruments that he had not made but which appeared to copy his. We also showed him the effects of the contact of different metals on the nerves of frogs. The names of Galvani and Volta had not yet echoed in these vast solitudes.

After the electric apparatus, made by a clever in-habitant of the llanos, nothing interested us more in Calabozo than the gymnoti, living electric apparatuses. I had busied myself daily over many years with the phenomenon of Galvanic electricity and had enthusi-astically experimented without knowing what I had discovered; I had built real batteries by placing metal discs on top of each other and alternating them with bits of muscle flesh, or other humid matter, and so was eager, after arriving at Cumaná, to obtain electric eels. We had often been promised them, and had always been deceived. Money means less the further from the coast you go, and there was no way to shake the imperturbable apathy of the people when even money meant nothing!

Under the name of *tembladores* ('which make you tremble') Spaniards confuse all electric fish. There are some in the Caribbean Sea, off the Cumaná coast. The

Guaiquerí Indians, the cleverest fishermen in the area, brought us a fish that numbed their hands. This fish swims up the little Manzanares river. It was a new species of ray whose lateral spots are hard to see, and which resembles Galvani's torpedo. The Cumaná torpedo was very lively, and energetic in its muscular contractions, yet its electric charges were weak. They became stronger when we galvanized the animal in contact with zinc and gold. Other *tembladores*, proper electric eels, live in the Colorado and Guarapiche rivers and several little streams crossing the Chaima Indian missions. There are many of them in the great South American rivers, the Orinoco, Amazon and Meta, but the strength of the currents and the depths prevent Indians from catching them. They see these fish less often than they feel their electric shocks when they swim in the rivers. But it is in the llanos, especially around Calabozo, between the small farm of Morichal and the *missions de arriba* and *de abaxo*, that the stagnant ponds and tributaries of the Orinoco are filled with electric eels. We wanted first to experiment in the house we lived in at Calabozo but the fear of the eel's electric shock is so exaggerated that for three days nobody would fish any out for us, despite our promising the Indians two piastres for each one. Yet they tell whites that they can touch *tembladores* without shock if they are chewing tobacco.

Impatient of waiting, and having only obtained uncertain results from a living eel brought to us, we went to the Caño de Bera to experiment on the water's edge. Early in the morning on the 19th of March we left for the little village of Rastro de Abaxo: from there Indians led us to a stream, which in the dry season forms a muddy pond surrounded by trees, clusia, amyris and mimosa with fragrant flowers. Fishing eels with nets is very difficult because of the extreme agility with which they dive into the mud, like snakes. We did not want to use *barbasco*, made with roots of *Piscidia erythrina*, *Jacquinia armillaris* and other species of phyllanthus which, chucked into the pond, numbs fish. This would have weakened the eel. The Indians decided to fish with their horses, *embarbascar con caballos*. It was hard to imagine this way of fishing; but soon we saw our guides returning from the savannah with a troop of wild horses and mules. There were about thirty of them, and they forced them into the water.

The extraordinary noise made by the stamping of the horses made the fish jump out of the mud and attack. These livid, yellow eels, like great water snakes, swim on the water's surface and squeeze under the bellies of the horses and mules. A fight between such different animals is a picturesque scene. With harpoons and long pointed reeds the Indians tightly circled the pond; some climbed trees whose branches hung over the water's surface. Screaming and prodding with their reeds they stopped the horses leaving the pond. The eels, dazed by the noise, defended themselves with their electrical charges. For a while it seemed they might win. Several horses collapsed from the shocks received on their most vital organs, and drowned under the water. Others, panting, their manes erect, their eyes anguished, stood up and tried to escape the storm surprising them in the water. They were pushed back by the Indians, but a few managed to escape to the bank, stumbling at each step, falling on to the sand exhausted and numbed from the electric shocks.

In less than two minutes two horses had drowned. The eel is about 5 feet long and presses all its length along the belly of the horse, giving it electric shocks. They attack the heart, intestines and the *plexus coeliacus* of the abdominal nerves. It is obvious that the shock felt by the horse is worse than that felt by a man touched on one small part. But the horses were probably not killed, just stunned. They drowned because they could not escape from among the other horses and eels.

We were sure that the fishing would end with the death of all the animals used. But gradually the violence of the unequal combat died down, and the tired eels dispersed. They need a long rest and plenty of food to recuperate the lost galvanic energy. The mules and horses seemed less frightened; their manes did not stand on end, and their eyes seemed less terrified. The eels timidly approached the shore of the marshy pond where we fished them with harpoons tied to long strings. While the string is dry the Indians do not feel any shocks. In a few minutes we had five huge eels, only slightly wounded. Later, more were caught.

The water temperature where these animals live is 26°c to 27°c. We are assured that their electric energy decreases in colder water. It is remarkable that these animals with electromotive organs are found not in the air but in a fluid that conducts electricity.

The eel is the largest of the electric fish; I have measured one that is 5 feet 3 inches long. Indians say they have seen even longer. A fish 3 feet 6 inches weighed 12 pounds. The eels from the Caño de Bera are of a pretty olive green, with a yellow mixed with red under their heads. Two rows of small yellow stains are placed symmetrically along their backs from the head to the tail. Each stain has an excretory opening. The skin is constantly covered with a mucus, which, as Volta has shown, conducts electricity twenty to thirty times more efficiently than pure water. It is odd that none of the electric fish discovered here are covered in scales.

It would be dangerous to expose yourself to the first shocks from a large excited eel. If by chance you get a shock before the fish is wounded, or exhausted by a long chase, the pain and numbness are so extreme that it is hard to describe the nature of the sensation. I do not remember ever getting such shocks from a Leyden jar as when I mistakenly stepped on a gymnotus just taken out of the water. All day I felt strong pain in my knees and in all my joints. Torpedoes and electric eels cause a twitching of the tendon in the muscle touched by the electric organ, which reaches one's elbow. With each stroke you feel an internal vibration that lasts two or three seconds, followed by a painful numbness. In the graphic language of the Tamanac Indians the electric eel is called *arimna*, which means 'something that deprives you of movement'.

While European naturalists find electric eels extremely interesting, the Indians hate and fear them.

However, their flesh is not bad, although most of the body consists of the electric apparatus, which is slimy and disagreeable to eat. The scarcity of fish in the marshes and ponds on the llanos is blamed on the eels. They kill far more than they eat, and Indians told us that when they capture young alligators and electric eels in their tough nets the eels do not appear to be hurt because they paralyse the young alligators before they themselves can be attacked. All the inhabitants of the waters flee the eels. Lizards, turtles and frogs seek ponds free of eels. At Uritici a road had to be redirected as so many mules were being killed by eels as they forded a river.

On the 24th of March we left Calabozo. At about four in the afternoon we found a young naked Indian girl stretched out on her back in the savannah; she seemed to be around twelve or thirteen. She was exhausted with fatigue and thirst, with her eyes, nose and mouth full of sand, and breathing with a rattle in her throat. Next to her there was a jar on its side, half full of sand. Luckily we had a mule carrying water. We revived her by washing her face and making her drink some wine. She was scared when she found herself surrounded by so many people, but she slowly relaxed and talked to our guides. From the position of the sun she reckoned she had fainted and remained unconscious for several hours. Nothing could persuade her to mount one of our mules. She wanted to return to Uritici where she had been a servant on a hacienda whose owner had sacked her after she had suffered a long illness because she could not work as well as before. Our threats and requests were useless; she was hardened to suffering, like all of her race, and lived in the present without fear of the future. She insisted on going to one of the Indian missions near Calabozo. We emptied her jar of sand and filled it with water. Before we had mounted our mules she had set off, and was soon a cloud of sand in the distance.

During the night we forded the Uritici river, home of numerous voracious alligators. We were told that we should not let our dogs drink from the river as alligators often leave the banks and chase dogs. We were shown a hut, or a kind of shed, where our host in Calabozo had had an extraordinary adventure. He was sleeping with a friend on a bench, covered with skins, when at dawn he was woken by a noise and violent shaking. Bits of earth flew about the hut, and suddenly a young alligator climbed up from under their bed and tried to attack a dog sleeping in the doorway; but it could not catch it, and ran to the bank and dived into the water. When they examined the ground under their bed they found it excavated; it was hardened mud where the alligator had spent its summer asleep, as they all do in the llano dry season. The noise of the men and horses, and the smell of dogs, had woken it up. The Indians often find enormous boas, which they call *uji*, or water snakes, in a similar state of lethargy. To revive them they sprinkle the boas with water. They kill them and hang them in a stream, and after they have rotted they make guitar strings from the tendons on their dorsal muscles, which are far better than strings made from howler-monkey guts.

Journey up the Apure River

Only after Diamante do you enter territory inhabited by tigers, crocodiles and *chiguires*, a large species of Linnaeus's genus *Cavia* (capybara). We saw flocks of birds pressed against each other flash across the sky like a black cloud changing shape all the time. The river slowly grew wider. One of the banks is usually arid and sandy due to flooding. The other is higher, covered with full-grown trees. Sometimes the river is lined with jungle on both sides and becomes a straight canal some 150 toises wide. The arrangement of the trees is remarkable. First you see the *sauso* shrubs (*Hermesia castaneifolia*), a hedge some 4 feet high as if cut by man. Behind this hedge a brushwood of cedar, Brazil-wood and *gayac*. Palms are rare; you see only scattered trunks of corozo and thorny *piritu*. The large quadrupeds of these regions, tigers, tapirs and peccaries, have opened passages in the *sauso* hedge. They appear through these gaps to drink water. They are not frightened of the canoes, so we see them skirting the river until they disappear into the jungle through a gap in the hedge. I confess that these often repeated scenes greatly appeal to me. The pleasure comes not solely from the curiosity a naturalist feels for the objects of his studies, but also to a feeling common to all men brought up in the customs of civilization. You find yourself in a new world, in a wild, untamed nature. Sometimes it is a jaguar, the beautiful American panther, on the banks; sometimes it is the hocco (*Crax alector*) with its black feathers and tufted head, slowly strolling along the *sauso* hedge. All kinds of animals appear, one after the other. 'Es como en el paraíso' ('It is like paradise') our old Indian pilot said. Everything here reminds you of that state of the ancient world revealed in venerable traditions about the innocence and happiness of all people; but when carefully observing the relationships between the animals you see how they avoid and fear each other. The golden age has ended. In this paradise of American jungles, as everywhere else, a long, sad experience has taught all living beings that gentleness is rarely linked to might.

Where the shore is very wide, the line of *sausos* remains far from the river. In the intermediate zone up to ten crocodiles can be seen stretched out in the sand. Immobile, with their jaws wide open at right angles, they lie next to each other without the least sign of sociability, unlike those animals that live in groups. The troop separates as soon as it leaves the shore; however, it consists probably of one male and numerous females as males are rare due to the rutting season when they fight and kill each other. There were so many of these great reptiles that all along the river we could always see at least five or six of them, although the fact that the Apure had not yet flooded meant that hundreds more of these saurians remained buried in the savannah's mud. The Indians told us that not a year went by without two or three people, mainly women going to fetch water, being torn apart by these carnivorous lizards. They told us the story of an Indian girl from Uritici who by her intrepidity and presence of mind had saved herself from the jaws of one of those monsters. As soon as she felt herself seized she poked her fingers so violently into the animal's eyes that pain forced it to drop its prey after slicing off one arm. Despite the copious bleeding the little Indian girl swam ashore with her remaining arm. In those lonely places where man lives in constant struggle with nature he must resort to any means to fight off a jaguar, a boa (*tragavenado*) or a crocodile; everyone is prepared for some sort of danger. 'I knew,' said the young Indian girl coolly, 'that the crocodile would let go when I stuck my fingers in its eyes.'

The Apure crocodiles find enough food eating *chiguires* (*Cavia capybara*), which live in herds of fifty to sixty on the river banks. These unhappy animals, as big as our pigs, cannot defend themselves. They swim better than they run, but in the water they fall to crocodiles and on land to jaguars. It is difficult to understand how, exposed to such formidable enemies, they remain so numerous; it can be explained only by how quickly they reproduce.

We stopped by the mouth of the Caño de la Triguera, in a bay called Vuelta de Jobal, to measure the speed of the current, which was 2.56 feet an hour. We were again surrounded by *chiguires*, who swim like dogs with their heads and necks out of the water. On the beach opposite we saw an enormous crocodile sleeping among these rodents. It woke when we approached and slowly slipped into the water without disturbing the rodents. Indians say that their indifference is due to stupidity, but perhaps they know that the Apure and Orinoco crocodile does not attack prey on land, only when it comes across one in its way as it slips into the water.

Near the Jobal nature takes on a more imposing and wild character. It was there we saw the largest tiger we have ever seen. Even the Indians were surprised by its prodigious size; it was bigger than all the Indian tigers I have seen in European zoos. The animal lay under the shade of a great zamang. It had just killed a *chiguire*; but it had not yet touched its victim, over which it rested a paw. *Zamuros*, a kind of vulture, had gathered in flocks to devour what was left over from the jaguar's meal. It was a strange scene, mixing daring with timidity. They hop to within 2 feet of the jaguar, but the slightest movement and they rush away. To observe their movements more closely we climbed into the small canoe accompanying our pirogue. It is very rare for a tiger to attack a canoe by swimming out to it, and it will only do this if it has been deprived of all food for a long time. The noise of our oars made the animal slowly get up and hide behind the *sauso* shrub along the bank. The vultures wanted to profit from this momentary absence to devour the dead *chiguire*. But the tiger, despite our proximity, jumped into their midst; and in a fit of anger, expressed by the movement of its tail, dragged his prey into the jungle. The Indians lamented not having their lances to leap ashore and attack the tiger with. They are used to this weapon and did not trust our rifles, which in the humidity refused to fire.

Going on down the river we met the large herd of *chiguires* that the tiger had scattered after choosing his victim. These animals watched us come ashore without panicking. Some sat and stared at us like rabbits, moving their upper lips. They did not seem to fear man, but the sight of our big dog put them to flight. As their hindquarters are higher than their front they run with little gallops but so slowly that we managed to capture two of them. The *chiguire*, who swims well, lets out little groans when it runs as if it had difficulty breathing. It is the largest of the rodents; it defends itself only when it is surrounded or wounded. As its grinding teeth, particularly those at the back, are very strong and long it can, simply by biting, tear the paw off a tiger or the leg off a horse. Its flesh smells disagreeably of musk, yet local ham is made from it. That explains the name, 'water pig', given to it by ancient naturalists. Missionary monks do not hesitate to eat this ham during Lent. According to their classification they place the armadillo, the *chiguire* and the manatee near the turtles; the first because it is covered with a hard armour, like a kind of shell; the other two because they are amphibious. On the banks of the Santo Domingo, Apure and Arauca rivers, and in the marshes of the flooded savannahs, the *chiguires* reach such numbers that the pasture lands suffer. They graze the grass that fattens the horses, called *chiguirero*. They also eat fish.

These animals, scared by the approach of our boat, stayed for eight to ten minutes under water.

We spent the night as usual in the open air, though we were in a plantation whose owner was hunting tigers. He was almost naked, and brownish-black like a *zambo*; this did not stop him believing that he was from the caste of whites. He called his wife and daughter, as naked as he was, Doña Isabela and Doña Manuela. Without ever having left the Apure river bank they took a lively interest 'in news from Madrid, of the unending wars, and that kind of thing from over there (*todas las cosas de allá*)'. He knew that the King of Spain would soon come and visit 'the great Caracas country', yet he added pleasingly, 'As people from the court eat only wheat bread they would never go beyond the town of Victoria so we would never see them here.' I had brought a *chiguire* with me that I wanted to roast; but our host assured me that '*Nosotros caballeros blancos* (white men like he and I) were not born to eat Indian game.' He offered us venison, killed the evening before with an arrow, as he did not have powder or firearms.

We supposed that a small wood of banana trees hid the farm hut; but this man, so proud of his nobility and the colour of his skin, had not bothered to build even an *ajupa* with palm leaves. He invited us to hang our hammocks near his, between two trees; and promised us, in a satisfied way, that if we returned up river during the rainy season we would find him under a roof. We would soon be complaining of a philosophy that rewards laziness and makes man indifferent to life's comforts. A furious wind rose after midnight, lightning crossed the sky, thunder groaned, and we were soaked to the bone. During the storm a bizarre accident cheered us up. Doña Isabela's cat was perched on the tamarind tree under which we were spending the night. He let himself fall into the hammock of one of our companions who, wounded by the cat's claws and woken from deepest sleep, thought he had been attacked by a wild animal. We ran up to him while he was screaming, and with embarrassment told him of his confusion. While the rain poured down on our hammocks and instruments Don Ignacio congratulated us on the good fortune of not having slept on the beach but on his land with well-bred white people, 'entre gente blanca y de trato'. As we were soaked it was hard to convince ourselves of this better situation, and we listened impatiently to the long story that our host told of his expedition to the Meta river, the bravery he had displayed in a bloody battle with the Guahibo Indians, and of the 'favours he had rendered to God and his King in kidnapping children (*los indiecitos*) from their parents to distribute them around the missions'. What an odd experience it was to find ourselves in these vast solitudes with a man who believed he was European, with all the vain pretensions, hereditary prejudices and mistakes of civilization, but whose only roof was a tree.

April 1st. At sunrise we said goodbye to Don Ignacio and Doña Isabela, his wife. We passed a low island where thousands of flamingoes, pink pelicans, herons and moorhens nested, displaying the most varied colours. These birds were so packed together that they gave the impression that they could not move. The island was called Isla de Aves.

We stopped on the right bank in a small mission inhabited by Indians of the Guamo tribe. There were some eighteen to twenty huts made of palm leaves, but in the statistics sent annually to the court by the missionaries this grouping of huts was registered under the name Pueblo de Santa Bárbara de Arichuna. The Guamo tribe refuse to be tamed and become sedentary. Their customs have much in common with the Achagua, Guahibo and Otomac, especially their dirtiness, their love for vengeance and their nomadic life-style, but their languages are completely different. These four tribes live principally from fishing and hunting on the often flooded plains between the Apure, Meta and Guaviare rivers. Nomadic life has been imposed by the physical conditions. The Indians of the Santa Bárbara mission could not offer us supplies as they grow only a little cassava. However, they were friendly, and when we entered their huts they gave us dried fish and water kept in porous jars where it stayed fresh.

We spent the night on a dry, wide beach. The night was silent and calm and the moon shone marvellously. The crocodiles lay on the beach so that they could see our fire. We thought that maybe the glow of the fire attracted them, as it did fish, crayfish and other water creatures. The Indians showed us tracks in the sand from three jaguars, two of them young; doubtless a female with cubs come to drink water. Finding no trees on the beach we stuck our oars in the sand and hung our hammocks. All was peaceful until about eleven when a dreadful noise began in the jungle around us that made sleep impossible. Among the many noises of screeching animals the Indians could recognize only those that were heard separately; the fluted notes of the *apajous*, the sighs of the alouate apes, the roar of the jaguar and puma; the calls of the peccary, sloth, hocco, *parraka* and other gallinaceous birds. When the jaguars approached the edge of the jungle our dog, who up to then had been barking continuously, began to growl and hid under our hammocks. Sometimes, after a long silence, we again heard the tiger's roar from the tops of trees, and then the din of monkeys' whistles as they fled from danger.

The confidence of the Indians helped to make us feel braver. One agrees with them that tigers fear fire, and never attack a man in his hammock. If you ask the Indians why jungle animals make such a din at certain moments of the night they say, 'They are celebrating the full moon.' I think that the din comes from deep in the jungle because a desperate fight is taking place. Jaguars, for example, hunt tapir and peccaries, who protect themselves in large herds, trampling down the vegetation in their way.

April 3rd. Since leaving San Fernando we had not met one boat on the beautiful river. Everything suggested the most profound solitude. In the morning the Indians had caught with a hook the fish called *caribe* [piranha] or *caribito* locally as no other fish is more avid for blood. It attacks bathers and swimmers by biting large chunks of flesh out of them. When one is slightly wounded it is difficult to leave the water without getting more wounds. Indians are terrified of the *caribe* fish and several showed us wounds on their calves and thighs, deep scars made by these little fish that the Maypure call *umati*. They live at the bottom of rivers, but as soon as a few drops of blood are spilled in the water they reach the surface in their thousands. When you consider the numbers of these fish, of which the most voracious and cruel are but 4 to 5 inches long, the triangular shape of their sharp, cutting teeth, and the width of their retractile mouths you cannot doubt the fear that the *caribe* inspires in the river inhabitants. In places on the river when the water was clear and no fish could be seen we threw bits of bloodied meat in, and within minutes a cloud of *caribes* came to fight for their food. I described and drew this fish on the spot. The *caribito* has a very agreeable taste. As one does not dare bathe when it is around you can regard it as the greatest scourge of this climate where mosquito bites and skin irritation make a bath so necessary.

At midday we stopped at a deserted spot called Algodonal. I left my companions while they beached the boat and prepared the meal. I walked along the beach to observe a group of crocodiles asleep in the sun, their tails, covered with broad scaly plates, resting on each other. Small herons, as white as snow, walked on their backs, even on their heads, as if they were tree trunks. The crocodiles were grey-green, their bodies were half covered in dried mud. From their colour and immobility they looked like bronze statues. However, my stroll almost cost me my life. I had been constantly looking towards the river, and then, on seeing a flash of mica in the sand, I also spotted fresh jaguar tracks, easily recognizable by their shape. The animal had gone off into the jungle, and as I looked in that direction I saw it lying down under the thick foliage of a ceiba, eighty steps away from me. Never has a tiger seemed so enormous.

There are moments in life when it is useless to call on reason. I was very scared. However, I was sufficiently in control of myself to remember what the Indians had advised us to do in such circumstances. I carried on walking, without breaking into a run or moving my arms, and thought I noted that the wild beast had its eye on a herd of capybaras swimming in the river. The further away I got the more I quickened my pace. I was so tempted to turn round and see if the cat was chasing me! Luckily I resisted this impulse, and the tiger remained lying down. These enormous cats with spotted skins are so well fed in this country well stocked with capybara, peccaries and deer that they rarely attack humans. I reached the launch panting and told my adventure story to the Indians, who did not give it much importance.

In the evening we passed the mouth of the Caño del Manati, named after the immense amount of manatees caught there every year. This herbivorous animal of the Cetacea family is called by the Indians *apcia* and *avia*, and reaches 10 to 12 feet long. The manatee is plentiful in the Orinoco. We dissected one that was 9 feet long while at Carichana, an Orinocan mission. The manatee eats so much grass that we found its stomach, divided into several cavities, and its intestines (108 feet long) filled with it. Its flesh is very savoury, though some prejudice considers it to be unwholesome and fever-producing. Its flesh when dried can last for a year. The clergy consider this mammal a fish, so they eat it at Lent.

We spent the night on Isla Conserva. The Indians had lit a campfire near the water. Again we confirmed that the glow of the flames attracted crocodiles and dolphins (*toninas*) whose noise stopped us sleeping until we decided to put the fire out. That night we had to get up twice. I mention this as an example of what it means to live in the jungle. A female jaguar approached our camp-site when it brought a cub to drink water. The Indians chased it away, but we heard its cub's cries, like a cat's, for hours. A little later our large dog was bitten, or stung as the Indians say, by some enormous bats that flew around our hammocks. The wound in its snout was tiny and the dog howled more from fear than pain.

April 4th. It was our last day on the Apure river. During several days a plague of insects had been torturing our hands and faces. They were not mosquitoes but *zancudos*, which are really gnats. They appear after sunset; their proboscises are so long that they can pierce your hammock, your canvas and your clothes from the other side.

Mouth of the Anaveni River – Uniana peak – Atures mission – Cataract or raudal of Mapara – Islets

As the Orinoco runs from south to north it crosses a chain of granite mountains. Twice checked in its course the river breaks furiously against rocks that form steps and transversal dykes. Nothing can be grander than this countryside. Neither the Tequendama Falls near Bogotá, nor the magnificent cordilleras surpassed my first impressions of the Atures and Maypures rapids. Standing in a position that dominates the uninterrupted series of cataracts it is as if the river, lit by the setting sun, hangs above its bed like an immense sheet of foam and vapours. The two great and famous cataracts of the Orinoco are formed as the river breaks through the Parima mountains. Indians call them Mapara and Quituna, but missionaries have substituted these names with Atures and Maypures, named after tribes living in two villages near by. On the Caracas coast the two Great Cataracts are simply called the two Raudales, from the Spanish *raudo*, 'rushing'.

Beyond the Great Cataracts an unknown land begins. This partly mountainous and partly flat land receives tributaries from both the Amazon and the Orinoco. No missionary writing about the Orinoco before me has passed beyond the Maypures *raudal*. Up river, along the Orinoco for a stretch of over 100 leagues, we came across only three Christian settlements with some six to eight whites of European origin there. Not surprisingly, such a deserted territory has become the classic place for legends and fantastic histories. Up here serious missionaries have located tribes whose people have one eye in the middle of their foreheads, the heads of dogs, and mouths below their stomachs. It would be wrong to attribute these exaggerated fictions to the inventions of simple missionaries because they usually come by them from Indian legends. From his vocation, a missionary does not tend towards scepticism; he imprints on his memory all that the Indians have repeated and when back in Europe delights in astonishing people by reciting facts he has collected. These travellers' and monks' tales (*cuentos de viajeros y frailes*) increase in improbability the further you go from the Orinoco forests towards the coasts inhabited by whites. When at Cumaná you betray signs of incredulity, you are silenced by these words, 'The fathers have seen it, but far above the Great Cataracts *más arriba de los Raudales.*'

April 15th. At dawn we passed the Anaveni river, a tributary river that comes down from mountains in the east. The heat was so excessive that we stayed for a long time in a shaded place, fishing, but we could not carry off all the fish we hooked. Much later we reached the foot of the Great Cataract in a bay, and took the difficult path – it was night by then – to the Atures mission, a league away. We found this mission in a deplorable state. At the time of Solano's boundary expedition it contained 320 Indians. Today it has only forty-seven. When it was founded Atures, Maypures, Meyepures, Abanis and Uirupas tribes lived there, but now there were only Guahibos and a few families from the Macos left. The Atures have completely disappeared; the little known about them comes from burial caves in Ataruipe.

Between the 4th and 8th degrees of latitude the Orinoco not only divides the great jungles of Parima from the bare savannahs of the Apure, Meta and Guaviare but also separates tribes of very different customs. On the west the Guahibo, Chiricoa and Guamo tribes wander through treeless plains. They are filthy, proud of their independence, wild, and hard to settle in a fixed place to do regular work. Spanish missionaries call them *indios andantes* (wandering Indians). On the east of the Orinoco live the Maco, Saliva, Curacicana and Pareca tribes; they are tame, peaceful farmers who easily adapt to missionary discipline.

In the Atures mission both types of tribe can be found. With the missionary we visited huts of the Macos. The independent Macos are orderly and clean. They have their *rochelas*, or villages, two or three days' journey away. They are very numerous and like all wild Indians cultivate cassava, not maize. Thanks to these peaceful relations some *indios monteros*, or nomadic Indians, had established themselves in the mission a short time before. They insistently asked for knives, hooks and coloured glass beads, which they sewed on to their *guayucos* (*perizomas*). Once they got what they wanted they slipped back to the jungle because missionary discipline was not to their liking. Epidemics of fever, so common at the start of the rainy season, contributed to the desertion. Jungle Indians have a horror of the life of civilized man and desert when the slightest misfortune befalls them in the mission.

Smallpox, which has so devastated other areas of America that Indians burn their huts, kill their children and avoid any grouping of tribes, is not one of the reasons for the depopulation of the Raudales. In the Upper Orinoco this plague is almost unknown. Desertion from Christian missionaries must be sought more in the Indian's hate for the discipline, the poor food, the awful climate, and the unpardonable custom that Indian mothers have of using poisonous herbs to avoid pregnancy. Many of the women do not want to have babies. If they do have children they are not only exposed to jungle dangers but also to absurd superstitions. When twins are born family honour demands that one be killed. Indians say: 'To bring twins into the world is to be exposed to public scorn, it is to resemble rats, sarigues and the vilest animals.' And, 'Two children born at the same time cannot belong to the same father.' If a newborn child shows some physical deformity the father kills it immediately. They want only well-formed, robust children because deformities indicate some evil spell. Among the Orinoco Indians the father returns home only to eat or sleep in his hammock; he shows no affection for his children or his wife, who are there only to serve him.

While we unloaded the pirogue we investigated the impressive spectacle of a great river squeezed and reduced to foam. Instead of just describing my own sensations I shall try to paint an overall view of one of the most famous spots in the New World. The more imposing and majestic a scene, the more important it is to capture it in its smallest details, to fix the outline of the picture that you want to present to the reader's imagination, and to simply describe the particular characteristics of the great monuments of nature.

Throughout his entire journey through the Lower Orinoco the traveller faces only one danger: the natural rafts formed by drifting trees uprooted by the river. Woe to the canoes that at night strike one of these rafts of tangled lianas and tree trunks! When Indians wish to attack an enemy by surprise they tie several canoes together and cover them with grass to make it seem like a tangle of trees. Today Spanish smugglers do the same to avoid customs in Angostura.

Above the Anaveni river, between the Uniana and Sipapu mountains, you reach the Mapara and Quituna cataracts, commonly called by missionaries the Raudales. These natural weirs crossing from one side to the other offer the same picture: one of the greatest rivers in the world breaks into foam among many islands, rocky dykes and piles of granite blocks covered in palms.

April 16th. Towards evening we heard that our boats had passed both rapids in less than six hours, and arrived in good condition at the Puerto de Arriba. 'Your boat will not be wrecked because you are not carrying goods, and you travel with the monk of the Raudales,' a little brown man said to us bitterly. By his accent we recognized him as a Catalan. He traded in tortoise oil with the mission Indians, and was not a friend of the missionaries. 'The frail boats belong to us Catalans who, with permission from the Guianan Government, but not from the president of the mission, try to trade above the Atures and Maypures. Our boats are wrecked in the Raudales, key to all the missions beyond, and then Indians take us back to Carichana and try to force us to stop trading.' What is the source of this deep hatred of the missions in the Spanish colonies? It cannot be because they are rich in the Upper Orinoco. They have no houses, no goats and few cows. The resentment is aimed at the ways the missionaries obstinately close their territories off to white men.

In the little Atures church we were shown remains of the Jesuits' wealth. A heavy silver lamp lay half buried in sand. This object did not tempt the Indians; the Orinoco natives are not thieves, and have a great respect for property. They do not even steal food, hooks or axes. At Maypures and Atures locks on doors are unknown.

The missionary told us a story about the jaguars. Some months before our arrival a young jaguar had wounded a child while playing with him. I have verified the facts on the spot; it should interest those who study animal behaviour. Two Indian children, a boy and a girl of about eight and nine years of age, were sitting on the grass near the village when a jaguar came out from the jungle and ran round the children, jumping and hiding in the high grass, like our cats. The little boy sensed danger only when the jaguar struck him with its paw until blood began to flow. The little girl chased it off with branches from a tree. The intelligent little boy was brought to us. The jaguar's claw had ripped skin from his forehead. What did this playfulness mean in the jaguar? If the jaguar was not hungry, why did it approach the children? There is something mysterious in the sympathies and hatreds of animals.

In this area there are several species of peccaries, or pigs with lumbar glands, only two of which are known to naturalists in Europe. The Indians call the little peccary a *chacharo*. Reared in their houses they become tame like our sheep and goats. Another kind is called the *apida*, which is also domesticated and wanders in large herds. These animals announce themselves from a long way off because they break down all the shrubs in their way. During a botanical excursion Bonpland was warned by his Indian guides to hide behind a tree trunk as these *cochinos*, or *puercos del monte*, passed by. The flesh of the *chacharo* is flabby and disagreeable, but the Indians hunt them nevertheless, with small lances tied to cords. We were told at Atures that jaguars dread being surrounded by herds of wild pigs and climb trees to save themselves. Is this a hunters' tale, or a fact?

Among the monkeys we saw at the Atures mission we found one new species, which the creoles call *machis*. It is the *ouavapavi*, with grey hair and a bluish face. This little animal is as tame as it is ugly. Every day in the missionary courtyard it would grab a pig and sit on its back all day. We have also seen it riding a large cat brought up in Father Zea's house.

It was at the cataracts that we first heard talk about

the hairy man of the jungle, called *salvaje*, who rapes women, builds huts, and sometimes eats human flesh. Neither Indians nor missionaries doubt the existence of this man-shaped monkey, which terrifies them. Father Gili seriously related the story of a lady from San Carlos who praised the gentle character of the man of the jungle. She lived several years with him in great domestic harmony, and only asked hunters to bring her back home because she and her children (rather hairy also) 'were tired of living far from a church'. This legend, taken by missionaries, Spaniards and black Africans from descriptions of the orang-utang, followed us for the five years of our journey. We annoyed people everywhere by being suspicious of the presence of a great anthropomorphic ape in the Americas.

After two days near the Atures cataract we were happy to load the canoe again and leave a place where the temperature was usually 29°c by day and 26°c at night. All day we were horribly tormented by mosquitoes and *jejenes*, tiny venomous flies (or *simuliums*), and all night by *zancudos*, another kind of mosquito feared even by the Indians. Our hands began to swell, and this swelling increased until we reached the banks of the Temi. The means found to escape these insects are often quite original. The kind missionary Father Zea, all his life tormented by mosquitoes, had built a small room near his church, up on a scaffolding of palm trunks, where you could breathe more freely. At night we climbed up a ladder to dry our plants and write our diary. The missionary had correctly observed that the insects preferred the lower levels, that is, from the ground up to some 15 feet. At Maypures the Indians leave their villages at night and sleep near the cataracts because the mosquitoes seem to avoid air loaded with vapours.

Those who have not travelled the great rivers of tropical America, like the Orinoco or the Magdalena, cannot imagine how all day long, ceaselessly, you are tormented by mosquitoes that float in the air, and how this crowd of little animals can make huge stretches of land uninhabitable. However used to the pain you may become, without complaining; however much you try to observe the object you are studying, the mosquitoes, *jejenes* and *zancudos* will tear you away as they cover your head and hands, pricking you with their needle-like suckers through your clothes, and climbing into your nose and mouth, making you cough and sneeze whenever you try to talk. In the Orinoco missions the *plaga de las moscas*, or plague of mosquitoes, is an inexhaustible subject of conversation. When two people meet in the morning the first questions they ask each other are, 'Que le han parecido los zancudos de anoche?' and 'Como estamos hoy de mosquitos?' ('How were the *zancudos* last night?' and 'How are we for mosquitoes today?').

The lower strata of air, from the ground to some 20 feet up, are invaded by poisonous insects, like thick clouds. If you stand in a dark place, such as a cave formed by granite blocks in the cataracts, and look towards the sunlit opening you will see actual clouds of mosquitoes that get thicker or thinner according to the density of insects. I doubt that there is another country on earth where man suffers more cruelly during the rainy season than here. When you leave latitude 5 the biting lessens, but in the Upper Orinoco it becomes more painful because it is hotter, and there is absolutely no wind so your skin becomes more irritated. 'How good it would be to live on the moon,' a Saliva Indian said. 'It is so beautiful and clear that it must be free of mosquitoes.'

Whoever lives in this region, whether white, mulatto, black or Indian, suffers equally from insect stings. People spend their time complaining of the *plaga, del insufrible tormento de las moscas*. I have mentioned the curious fact that whites born in the Tropics can walk about barefoot in the same room where a recently arrived European runs the risk of being bitten by *niguas*, or chigoes (*Pulex penetrans*). These hardly visible animals dig under toenails and soon reach the size of a pea as they develop their eggs, situated in little sacs under their abdomens. It seems as if the *nigua* is able to distinguish the cellular membrane and blood of a European from those of a white *criollo*, something that the most detailed chemical analysis has been unable to do. It is not the same with mosquitoes, despite what is said on South American coasts. These insects attack Indians as much as Europeans; only the consequences of the bites vary with race. The same venomous liquid applied to the skin of a copper-coloured Indian and to a recently arrived white does not cause inflammations to the first, while to the second it causes hard, inflamed blisters that last for various numbers of days.

All day, even when rowing, Indians continually slap each other hard with the palm of the hand to scare off mosquitoes. Brusque in all their movements they continue to slap each other mechanically while they sleep. At Maypures we saw young Indians sitting in a circle, cruelly scratching each other's back with bark dried by the fire. With that patience only known in the copper-coloured race, some Indian women busied themselves by digging small lumps of coagulated blood from each bite with a sharp, pointed bone. One of the wildest Orinoco tribes, the Otomacs, use mosquito nets woven from fibre from the moriche palm. In villages on the Magdalena river Indians often invited us to lie down on oxhides near the church in the middle of the *plaza grande* where they had herded all the cattle, as the proximity of cattle gives you some respite from bites. When Indians saw that Bonpland was unable to prepare his plants because of the plague of mosquitoes they invited him into their 'ovens' (*hornitos*), as they call these small spaces without doors or windows, which they slide into on their bellies through a low opening. Thanks to a fire of greenwood, which gives off plenty of smoke, they expel all the insects and then block the 'oven' door. Bonpland, with a praiseworthy courage and patience, dried hundreds of plants shut up in these Indian *hornitos*.

The trouble an Indian takes to avoid the insects proves that despite his different skin colour he is just as sensitive to mosquito bites as any white. Irritability is increased by wearing warm clothes, by applying alcoholic liquors, by scratching the wounds, and – and this

I have observed myself – by taking too many baths. By bathing whenever we could Bonpland and I observed that a bath, though soothing for old bites, made us more sensitive to new ones. If you take a bath more than twice a day the skin becomes nervously excited in a way nobody in Europe could understand. It seems as if all one's sensitivity has become concentrated in the epidermic layers. Today the dangers that prevent Spaniards navigating up the Orinoco do not come from wild Indians or snakes or crocodiles or jaguars but, as they naïvely say, from 'el sudar y las moscas' (sweating and mosquitoes).

I have shown that winged insects which live in society and whose suckers contain a liquid that irritates skin make vast territories virtually uninhabitable. Other insects, just as small, called termites (*comején*) create insuperable obstacles to the progress of civilization in several hot countries. They rapidly devour paper, cardboard and parchment, and thus destroy archives and libraries. Whole provinces of Spanish America do not have any document that dates back more than a hundred years.

Garcita cataract – Maypures – Quituna cataract – Confluence of the Vichada and Zama – Aricagua rock – Siquita

Our boat was waiting for us in the Puerto de Arriba above the Atures cataract. On the narrow path that led to the *embarcadero* we were shown the distant rocks near the Ataruipe caves. We did not have time to visit that Indian cemetery though Father Zea had not stopped talking about the skeletons painted red with *onoto* inside the great jars. 'You will hardly believe,' said the missionary, 'that these skeletons and painted vases, which we thought unknown to the rest of the world, have brought me trouble. You know the misery I endure in the Raudales. Devoured by mosquitoes, and lacking in bananas and cassava, yet people in Caracas envy me! I was denounced by a white man for hiding treasure that had been abandoned in the caves when the Jesuits had to leave. I was ordered to appear in Caracas in person and journeyed pointlessly over 150 leagues to declare that the cave contained only human bones and dried bats. However, commissioners were appointed to come up here and investigate. We shall wait a long time for these commissioners. The cloud of mosquitoes (*nube de moscas*) in the Raudales is a good defence.'

April 17th. After walking for three hours we reached

our boat at about eleven in the morning. Father Zea packed provisions of clumps of bananas, cassava and chicken with our instruments. We found the river free of shoals, and after a few hours had passed the Garcita *raudal* whose rapids are easily crossed during high water. We were struck by a succession of great holes, more than 180 feet above the present water-level, that appeared to have been caused by water erosion. The night was clear and beautiful but the plague of mosquitoes near the ground was such that I was unable to record the level of the artificial horizon and lost the opportunity of observing the stars.

April 18th. We set off at three in the morning in order to reach the cataracts known as the Raudal de Guahibos before nightfall. We moored at the mouth of the Tomo river, and the Indians camped on the shore. At five in the afternoon we reached the *raudal*. It was extremely difficult to row against the current and the mass of water rushing over a bank several feet high. One Indian swam to a rock that divided the cataract in two, tied a rope to it, and began hauling our boat until, halfway up, we were able to get off with our instruments, dried plants and bare provisions. Surprisingly we found that above the natural wall over which the river fell there was a piece of dry land. Our position in the middle of the cataract was strange but without danger. Our companion, the missionary father, had one of his fever fits, and to relieve him we decided to make a refreshing drink. We had taken on board at Apures a *mapire*, or Indian basket, filled with sugar, lemons and grenadillas, or passion-fruit, which the

Spaniards call *parchas*. As we had no bowl in which to mix the juices we poured river water into one of the holes in the rock with a *tutuma*, and then added the sugar and acid fruit juices. In a few seconds we had a wonderfully refreshing juice, almost a luxury in this wild spot, but necessity had made us more and more ingenious. After quenching our thirst we wanted to have a swim. Carefully examining the narrow rocky dyke on which we sat, we saw that it formed little coves where the water was clear and still. We had the pleasure of a quiet bathe in the midst of noisy cataracts and screaming Indians. I enter into such detail to remind those who plan to travel afar that at any moment in life pleasures can be found.

After waiting for an hour we saw that our pirogue had safely crossed the *raudal*. We loaded our instruments and provisions and left the Guahibo rock. We began a journey that was quite dangerous. Above the cataract the river is some 800 toises wide and must be crossed obliquely at the point where the waters start rushing towards the fall. The men had been rowing for over twenty minutes when the pilot said that instead of advancing we were drifting back to the falls. Then there was a storm and heavy rain fell. Those anxious moments seemed to last for ever. The Indians whisper as they always do when in danger; but they rowed very hard and we reached the port of Maypures by nightfall.

The night was very dark and it would take us two hours to reach the village of Maypures. We were soaked to the skin, and after it stopped raining the *zancudos* returned. My companions were undecided as to whether to camp in the harbour or walk to the village. Father Zea insisted on going to the village where, with help from Indians, he had begun to build a two-floored house. 'You will find there,' he said naïvely, 'the same comforts as you have out of doors. There are no tables or chairs but you will suffer less from mosquitoes because in the mission they are not as shameless as down by the river.' We followed the missionary's advice. He ordered torches of copal to be lit. These are tubes of bark filled with copal resin. At first we passed beds of slippery rock, then a thick palm grove. We twice had to cross streams over tree trunks. The torches burned out. They give off more smoke than light, and easily extinguish. Our companion, Don Nicolas Soto, lost his balance in the dark crossing a marsh and fell off a tree trunk. For a while we had no idea how far he had fallen, but luckily it was not far and he was not hurt. The Indian pilot, who spoke Spanish quite well, did not stop saying how easy it would be to be attacked by snakes or jaguars. This is the obligatory topic of conversation when you travel at night with Indians. They think that by frightening European travellers they will become more necessary to them, and will win their confidence.

The Maypures cataracts appear like a cluster of little waterfalls following each other, as if falling down steps. The *raudal*, the name given by Spaniards to these kind of cataracts, is made up of a veritable archipelago of small islands and rocks which narrow the river so thoroughly that there is often less than 18 to 21 feet for boats to navigate through.

At the confluence of the Cameji and Orinoco we unloaded our baggage, and the Indians, familiar with all the shoals in the *raudal*, led the empty pirogue to the mouth of the Toparo river where the water is no longer dangerous. Each rock forming the falls of the *raudal* has a different name. As long as they are not more than 1.5 to 2 feet above water the Indians do not mind letting the current take their canoes; but to go up river they swim ahead and after much struggling tie cables to rocks and pull the boats up.

Sometimes, and it's the only accident the Indians fear, the canoes break against rocks. Then, their bodies bloodied, the Indians try to escape the whirlpools and swim ashore. In those places where the rocks are very high, or the embankment they are going up crosses the whole river, they roll the boat up on tree trunks.

The most famous cataracts, with the most obstacles, are called Purimarimi and Manimi and are about 3 metres wide. The difficulties involved in reaching these places, and the foul air filled with millions of mosquitoes, made it impossible to take a geodesical levelling, but with the aid of a barometer I was amazed that the whole fall of the *raudal* from the mouth of the Cameji to the Toparo was only some 27 to 30 feet. My surprise was related to the terrible din and foam flying from the river.

From the Manimi rock there is a marvellous view. Your eyes survey a foaming surface that stretches away for almost 2 leagues. In the middle of the waves rocks as black as iron, like ruined towers, rise up. Each island, each rock, is crowned by trees with many branches; a thick cloud floats above the mirror of the water and through it you see the tops of tall palms. What name shall we give these majestic plants? I guess that they are *vadgiai*, a new species, more than 80 feet high. Everywhere on the backs of the naked rocks during the rainy season the noisy waters have piled up islands of vegetation. Decorated with ferns and flowering plants these islands form flower-beds in the middle of exposed, desolate rocks. At the foot of the Manimi rock, where we had bathed the day before, the Indians killed a 7.5-foot snake, which we examined at leisure. The Macos called it a *camudu*. It was beautiful, and not poisonous. I thought at first that it was a boa, and then perhaps a python. I say 'perhaps' for a great naturalist like Cuvier appears to say that pythons belong to the Old World, and boas to the New. I shall not add to the confusions in zoological naming by proposing new changes, but shall observe that the missionaries and the latinized Indians of the mission clearly distinguish the (boa) from the *culebra de agua*, which is like the *camudu*.

In the time of the Jesuits the Maypures *raudal* mission was well known and had as many as 600 inhabitants including several families of whites. Under the government of the fathers of the Observance this has shrunk to some sixty. Those who still live there are mild and moderate, and very clean. Most of the wild Indians of the Orinoco are not excessively fond of strong alcohol like the North American Indian. It is true that Otomacs, Yaruros, Achaguas and Caribs often get drunk on *chicha* and other fermented drinks made from cassava, maize and sugared palm-tree fruit. But travellers, as usual, have generalized from the habits of a few villages. We often could not persuade the Guahibos who worked with us to drink brandy even when they seemed exhausted.

They grow banana and cassava, but not maize. Like the majority of Orinoco Indians, those in Maypures also make drinks that could be called nutritious. A famous one in the country is made from a palm called the *seje*, which grows wild in the vicinity. I estimated the number of flowers on one cluster at 44,000; the fruit that fall without ripening amount to 8,000. These fruit are little fleshy drupes. They are thrown into boiling water for a few minutes to separate the pulp, which has a sweet taste, from the skin, and are then pounded and bruised in a large vessel filled with water. Taken cold, the infusion is yellowish and tastes like almond milk. Sometimes *papelón* (unrefined sugar) or sugar cane is added. The missionary said that the Indians become visibly fatter during the two or three months when they drink this *seje* or dip their cassava cakes in it. The *piaches*, or Indian shamans, go into the jungle and sound the *botuto* (the sacred trumpet) under *seje* palm trees 'to force the tree to give a good harvest the following year'.

'Tengo en mi pueblo la fábrica de loza' (I have a pottery works in my village), Father Zea told us and led us to the hut of an Indian family who were baking large earthenware vessels, up to 2.5 feet high, out in the open on a fire of shrubs. This industry is characteristic of the diverse branches of the Maypures tribes, cultivated since time immemorial. Wherever you dig up the ground in the jungle, far from any human habitations, you find bits of painted pottery. It is noteworthy that the same motifs are used everywhere. The Maypures Indians painted decorations in front of us that were identical to those we had seen on the jars from the Ataruipe caves, with wavy lines, figures of crocodiles, monkeys and a large quadruped that I did not recognize but which was always crouched in the same position.

With the Maypures Indians, it is the women who decorate the vessels, cleaning the clay by washing it several times; then they shape it into cylinders and mould even the largest jars with their hands. The American Indians never discovered the potter's wheel.

It was fascinating to see *guacamayos*, or tame macaws, flying around the Indian huts as we see pigeons in Europe. This bird is the largest and most majestic of the parrot species. Including its tail it measures 2 feet 3 inches. The flesh, which is often eaten, is black and rather tough. These macaws, whose feathers shine with tints of purple, blue and yellow, are a grand ornament in Indian yards, and are just as beautiful as the peacock or golden pheasant. Rearing parrots was noticed by Columbus when he first discovered America.

Near the Maypures village grows an impressive tree some 60 feet high called by the colonists the *fruta de burro*. It is a new species of annona. The tree is famous for its aromatic fruit whose infusion is an efficient febrifuge. The poor missionaries of the Orinoco who suffer tertian fevers most of the year rarely travel without a little bag of *fruta de burro*.

April 21st. After spending two and a half days in the little village of Maypures near the Great Cataracts, we embarked in the canoe that the Carichana missionary had got for us. It had been damaged by the knocks it had received in the river, and by the Indians' carelessness. Once you have passed the Great Cataracts you feel you are in a new world; that you have stepped over the barriers that nature seems to have raised between the civilized coasts and the wild, unknown interior. On the way to the landing-stage we caught a new species of tree frog on the trunk of a hevea. It had a yellow belly, a back and head of velvety purple, and a narrow white stripe from its nose to its hind parts. This frog was 2 inches long; probably allied to the *Rana tinctoria* whose blood, so it is said, makes the feathers that have been plucked out of a parrot grow again in frizzled yellow and red if poured on to its skin.

The Indians told us that the jungles which cover the banks of the Sipapo abound in the climbing plant called *bejuco de maimure*. This species of liana is very important for the Indians as they weave baskets and mats from it. The Sipapo jungles are completely unknown. It is there that the missionaries place the Rayas tribe, whose mouths are said to be in their navels, perhaps due to the analogy with rays, whose mouths appear to be halfway down their bodies. An old Indian we met at Carichana, who boasted that he had often eaten human flesh, had seen these headless people 'with his own eyes'.

At the mouth of the Zama river we entered into a fluvial network worthy of attention. The Zama, Mataveni, Atabapo, Tuamini, Temi and Guianai rivers have *aguas negras*, that is, seen as a mass they appear brown like coffee, or greenish-black. These waters are, however, beautifully clear and very tasty. I have observed that crocodiles and mosquitoes, but not the *zancudos*, tend to avoid the black waters.

April 24th. A violent rainstorm forced us to embark before dawn. We left at two in the morning and had to abandon some books, which we could not find in the dark. The river runs straight from south to north, its banks are low and lined with thick jungle.

[. . .]

Humboldt reaches the Río Negro

May 6th. The morning was fresh and beautiful. For thirty-six days we had been locked up in a narrow canoe which was so unsteady that standing up suddenly from your seat would have capsized it. We had cruelly suffered from insect bites, but we had survived this unhealthy climate, and had crossed the many waterfalls and dykes that block the rivers and make the journey more dangerous than crossing the seas, without sinking. After all that we had endured, it gives me pleasure to speak of the joy we felt in having reached a tributary of the Amazon, of having passed the isthmus that separates the two great river systems. The uninhabited banks of the Casiquiare, covered in jungle, busied my imagination. In this interior of a new continent you get used to seeing man as not essential to the natural order. The earth is overloaded with vegetation: nothing prevents its development. An immense layer of mould manifests the uninterrupted action of organic forces. Crocodile and boa are the masters of the river; jaguar, peccary, the dante and monkeys cross the jungle without fear or danger, established there in an ancient heritage. This view of a living nature where man is nothing is both odd and sad. Here, in a fertile land, in an eternal greenness, you search in vain for traces of man; you feel you are carried into a different world from the one you were born into.

THE STORY OF PENGUIN CLASSICS

Before 1946 ...'Classics' are mainly the domain of academics and students, without readable editions for everyone else. This all changes when a little-known classicist, E. V. Rieu, presents Penguin founder Allen Lane with the translation of Homer's *Odyssey* that he has been working on and reading to his wife Nelly in his spare time.

1946 *The Odyssey* becomes the first Penguin Classic published, and promptly sells three million copies. Suddenly, classic books are no longer for the privileged few.

1950s Rieu, now series editor, turns to professional writers for the best modern, readable translations, including Dorothy L. Sayers's *Inferno* and Robert Graves's *The Twelve Caesars*, which revives the salacious original.

1960s The Classics are given the distinctive black jackets that have remained a constant throughout the series's various looks. Rieu retires in 1964, hailing the Penguin Classics list as 'the greatest educative force of the 20th century'.

1970s A new generation of translators arrives to swell the Penguin Classics ranks, and the list grows to encompass more philosophy, religion, science, history and politics.

1980s The Penguin American Library joins the Classics stable, with titles such as *The Last of the Mohicans* safeguarded. Penguin Classics now offers the most comprehensive library of world literature available.

1990s The launch of Penguin Audiobooks brings the classics to a listening audience for the first time, and in 1999 the launch of the Penguin Classics website takes them online to a larger global readership than ever before.

The 21st Century Penguin Classics are rejacketed for the first time in nearly twenty years. This world famous series now consists of more than 1300 titles, making the widest range of the best books ever written available to millions – and constantly redefining the meaning of what makes a 'classic'.

The Odyssey continues ...

The best books ever written

PENGUIN 🐧 CLASSICS

SINCE 1946

Find out more at www.penguinclassics.com